Nanobiotechnology in Food: Concepts, Applications and Perspectives

Hoda Jafarizadeh-Malmiri • Zahra Sayyar
Navideh Anarjan • Aydin Berenjian

Nanobiotechnology in Food: Concepts, Applications and Perspectives

Springer

Hoda Jafarizadeh-Malmiri
Faculty of Chemical Engineering,
East Azarbaijan
Sahand University of Technology
Tabriz, Iran

Zahra Sayyar
Faculty of Chemical Engineering,
East Azarbaijan
Sahand University of Technology
Tabriz, Iran

Navideh Anarjan
Faculty of Chemical Engineering,
East Azarbaijan
Islamic Azad University Tabriz Branch
Tabriz, Iran

Aydin Berenjian
Faculty of Engineering
The University of Waikato
Hamilton, Waikato, New Zealand

ISBN 978-3-030-05845-6 ISBN 978-3-030-05846-3 (eBook)
https://doi.org/10.1007/978-3-030-05846-3

Library of Congress Control Number: 2018966809

This Springer imprint is published by the registered company Springer Nature Switzerland AG
The registered company address is: Gewerbestrasse 11, 6330 Cham, Switzerland

Preface

Nanobiotechnology is defined as the convergence of nanotechnology and biology that is leading to the development of a new class of multifunctional devices and systems. Due to highly unique properties of nanomaterials, nanobiotechnology has gained a wide range of applications in fields such as biotechnology, chemical engineering, civil engineering, and medical sciences. In the food industry, however, not all scientists and engineers have recognized the potential applications of nanobiotechnology.

Distinctive chemical and physical properties of nanomaterials make them very attractive for new food product developments. During food processing, nanoparticles could be applied to improve nutritional quality, flow properties, sensory characteristics, and shelf life by decreasing the activity of the microorganisms. Indeed, nanobiotechnology might help in the development of healthier food with lower fat, sugar, and salts to overcome many food-related diseases. On the other hand, application of nanomaterials in food products may raise fears over their safety to consumer's health. Therefore, the presence of the nanoscale-sized structures in food materials requires critical analysis of any potentially adverse effects.

Nanobiotechnology in Food: Concepts, Applications and Perspectives is a comprehensive reference book containing exhaustive information on nanobiotechnology and the scope of its applications in food industries. The subsequent chapters in this book consider the basic principles in all aspects of nanobiotechnology and its key role in functional food product development. The key features of the book include challenges for nanobiotechnology, novel technologies in food nanobiotechnology, nano-additives for food industries, nanobiotechnology in food packaging, nano-sensors in food nanobiotechnology, and nano-encapsulation for targeted nutrition delivery. Subsequent chapters also cover the potential hazards of nanoparticles, commercialization consideration, and future prospects of nanobiotechnology in food industries.

The book is beneficial for graduate students, researchers, and scientists working on the application of nanobiotechnology to improving and facilitating the production of functional foods.

Tabriz, Iran — Hoda Jafarizadeh-Malmiri
Tabriz, Iran — Zahra Sayyar
Tabriz, Iran — Navideh Anarjan
Hamilton, New Zealand — Aydin Berenjian

Contents

1 Nanobiotechnology at a Glance 1
1.1 Introduction 1
1.2 Biotechnology Definition 2
1.2.1 Why Biotechnology? 3
1.2.2 Food Biotechnology 4
1.3 Nanotechnology Definition 8
1.3.1 Why Nanotechnology? 9
1.3.2 Food Nanotechnology 11
1.4 Nanobiotechnology Definition 12
1.4.1 Why Nanobiotechnology? 12
1.4.2 Food and Nanobiotechnology 13
References 16

2 Challenges for Nanobiotechnology 19
2.1 Introduction 19
2.2 Safety Aspects 20
2.3 Public Acceptance 21
2.4 Risk Assessment 23
2.5 Regulatory Aspects 24
References 24

3 Novel Technologies in Food Nanobiotechnology 27
3.1 Introduction 27
3.2 Enzyme Immobilization 27
3.3 Green Synthesis of Inorganic Nanoparticles 30
3.4 Nanoemulsions 36
References 39

4 Nano-additives for Food Industries 41
4.1 Introduction 41
4.2 Nanomaterial Classifications 42

4.2.1 Inorganic Nanomaterials 42
4.2.2 Surface Functionalized Nanomaterials 43
4.2.3 Organic Nanomaterials 44
4.2.4 Nanoencapsulated Compounds 44
4.3 Nanostructures' Food Additives 45
4.3.1 Solubility 46
4.3.2 Bioavailability 46
4.3.3 Control Release 50
4.3.4 Antioxidant Nanoparticles 51
4.3.5 Anti-browning Nanoparticles 52
4.3.6 Antimicrobial Nanoparticles 53
4.3.7 Mechanical Strength 55
4.3.8 Absorbent 56
4.3.9 UV-Blocker Nanoparticles 56
4.3.10 Pigmentation 57
4.3.11 Flavor 61
4.3.12 Nanosensors/Nanobiosensors 61
4.3.13 Other Active Nanoparticles 62
References 63

5 Nanobiotechnology in Food Packaging 69
5.1 Introduction 69
5.2 Active Packaging 70
5.3 Intelligent Packaging 72
5.4 Biodegradable Coatings and Films 73
5.5 Nanoparticles in Food Packaging 74
5.5.1 Organic Nanoparticles 74
5.5.2 Inorganic Nanoparticles 77
References 77

6 Nano-sensors in Food Nanobiotechnology 81
6.1 Introduction 81
6.2 Historical Developments of Bio Nanosensors 82
6.3 Classification of Biosensors 84
6.3.1 Electrochemical Biosensors 84
6.3.2 Potentiometric Biosensor 84
6.3.3 Calorimetric Biosensors 85
6.3.4 Amperometric Biosensor 85
6.3.5 Resonant Biosensors 85
6.3.6 Ion-Sensitive Biosensors 86
6.3.7 Optical Biosensors 86
6.4 Advantages and Types of Nanosensors in the Food Industry 86
6.4.1 Nanosensors Based on Nanostructure 87
6.4.2 Optical Nanosensors 88
6.4.3 Bio Nanosensors 88

6.4.4 Chemical Nanosensors ... 88
6.4.5 Physical Nanosensors ... 89
6.4.6 Sol-Gel Nanosensors ... 89
6.5 Nanosensors in the Agri-Food Industry ... 89
6.6 Mechanism of Biosensors ... 90
6.7 Nanosensor Applications ... 91
6.8 Disadvantage of Bio Nanosensors ... 91
References ... 94

7 Nano-encapsulation for Nutrition Delivery ... 95
7.1 Introduction ... 95
7.2 Nanoencapsulation ... 96
7.3 Materials for Nanoencapsulation ... 98
7.3.1 Polymer-Based Nanoencapsulation Materials ... 99
7.3.2 Lipid-Based Nanomaterials ... 100
7.3.3 Porous Inorganic Nanomaterials ... 100
7.3.4 Clay-Based Nanomaterials ... 100
7.4 Nanoencapsulation Techniques ... 101
7.4.1 Nanoprecipitation Method ... 102
7.4.2 Nanoemulsification Method ... 102
7.4.3 Coacervation Method ... 103
7.4.4 Spray Drying Method ... 105
7.4.5 Electrospinning and Electrospray Methods ... 107
7.4.6 Solvent Evaporation ... 108
7.4.7 Other Methods ... 109
7.5 Nanostructured Delivery Systems Applied in Encapsulation and Controlled Release ... 109
7.6 The Future of Controlled Release Systems ... 111
References ... 112

8 Potential Hazards of Nanoparticles ... 115
8.1 Introduction ... 115
8.2 Hazard and Risk Definition ... 117
8.3 Overall Theories ... 118
8.4 Critical Information of Physico-Chemical Properties and Evidence for Nanoparticle Toxicity ... 120
8.5 Hazard and Risk Evaluation for Nanoparticles ... 124
8.5.1 Risks of Inhaled Nanoparticles ... 126
8.5.2 Risks of Contacted Nanoparticles ... 126
8.5.3 Risks of Nanoparticles in the Aquatic Environment ... 126
8.6 Toxicity Pathways ... 127
8.7 Screening Hazards Test of Nanoparticle Applications ... 128
8.8 Adverse Outcome Pathways (AOP) ... 131
References ... 132

9 Commercialization Consideration 137
9.1 Introduction 137
9.2 The Purpose of Commercialization 138
9.3 The Process of Commercialization 138
9.4 Commercialization Strategies for Research, Patent and Nanotechnology Start-Ups 140
9.5 Market Entering Strategies and Potential Growth for Nano Products in Future 144
9.5.1 Medical Market 144
9.5.2 Food Market 145
9.6 Factors Influencing Nanotechnology Commercialization 146
9.7 Economic Impact of Nanotechnology Commercialization 147
9.8 Challenges and Barriers to Commercialization of Nanotechnology 147
9.9 Commercial Applications 148
References 150

10 Future Prospects of Nanobiotechnology 153
References 155

Chapter 1
Nanobiotechnology at a Glance

1.1 Introduction

Demand for more advanced technologies and innovations in useful materials and tools for the study of life are based on two technologies, namely biotechnology and nanotechnology. Biotechnology and nanotechnology are two of the twenty-first century's most promising technologies. Nanotechnology is established on the design, development and application of materials and devices possessing at least one dimension sized in a nanometer scale (Assa et al. 2016). As compared to the materials in their bulk form, nanomaterials have novel physical, chemical, mechanical, optical and biological properties due to their large surface to volume ratio, which is the key factor for the unique properties of the nanoparticles and nanomaterials (Haroon and Ghazanfar 2016; de Morais et al. 2014). This increases potential applications of the nanomaterials in the wide ranges of industries and products. On the other hand, biotechnology deals with metabolic and other physiological processes of biological subjects including living cells, microorganisms and enzymes. In fact, biotechnology uses the knowledge and techniques of biology to manipulate molecular, genetic and cellular processes to develop products and services that are used in diverse fields from medicine to agriculture. Nanobiotechnology integrates the design of new materials and devices with the exquisite specificity of biological molecules, enzymes and cells to solve critical problems in biology. In fact, nanobiotechnology is the combination of engineering and molecular biology (Raju 2016; Shoseyov and Levy 2008). In this chapter, the terms of biotechnology, nanotechnology and nanobiotechnology are defined. Furthermore, their importance and applications, especially in food areas, are presented.

H. Jafarizadeh-Malmiri et al., *Nanobiotechnology in Food: Concepts, Applications and Perspectives*, https://doi.org/10.1007/978-3-030-05846-3_1

1.2 Biotechnology Definition

Biotechnology is a field of applied biology that involves the use of living organisms and bioprocesses in engineering, technology, medicine and other fields requiring bio-products. Biotechnology also utilizes these products for manufacturing purposes. Modern use of similar terms include genetic and biochemical engineering as well as cell and tissue culture technologies. Biotechnology has different definitions according to numerous countries and organizations. The following definitions are provided for the term 'biotechnology':

- Biotechnology is the integration of natural sciences and engineering in order to achieve the application of organisms, cells, parts thereof and molecular analogues for products and services (The European Federation of Biotechnology).
- Biotechnology is the controlled use of biological agents, such as microorganisms or cellular components (US National Science Foundation).
- Biotechnology is any technique that uses living organisms or substances from these organisms, to make or modify a product to improve plants or animals or to develop microorganisms for specific uses (Office of Technology Assessment of United State Congress).
- Biotechnology is the application of biological organisms, systems, or processes by various industries to learn about the science of life and the improvement of the value of materials and organisms such as pharmaceuticals, crops, and livestock (American Chemical Society).
- Biotechnology is the application of biological organisms, system or process to manufacturing and service industries (British Biotechnologist).
- Biotechnology is technology using biological phenomena for copying and manufacturing various kinds of useful substances (Japanese Biotechnologists) (Elnashar 2010; Govil et al. 2017).

Biotechnology has been unwittingly used for several thousand years, initially in the area of food production and preservation as exemplified by the early production of alcoholic beverages and bread using microbial contaminants (Ibrahim and Day 2014). The name biotechnology was given by a Hungarian engineer, Karoly Ereky, in 1919 to describe a technology based on converting raw materials into a more useful product (Bud 1989).

Modern biotechnology is also referred to as genetic engineering, genetic modification or transgenic technology. In this technology, nuclear DNA is modified through insertion of gene (gene encoding desired trait). The modified DNA is called a recombinant DNA. When recombinant DNA expresses, it encodes the desired product. This technology, when implemented to enhance food qualities or yield, is called food technology. Modern biotechnology is helpful in enhancing taste, yield, shelf life and nutritive values. This technology is also useful in food processing (fermentation and enzyme involving processes). Therefore, biotechnology is beneficial in erasing hunger, malnutrition and diseases from developing and third world countries. Modern biotechnology products are commercially reasonable hence they can improve agriculture as well as food industry that will result in an increased income of poor farmers (Haroon and Ghazanfar 2016).

1.2.1 *Why Biotechnology?*

Biotechnology is roughly divided into four main general parts, namely green, red, white and blue biotechnology. Green biotechnology is a very important field of modern biotechnology. The foundation of green biotech is crop improvement and production of novel products in plants, which is achieved by implanting foreign genes to plant species that are economically important. This contains three main areas: plant tissue culture (i.e. rapid production of banana and citrus fruits), plant genetic engineering (i.e. creating improved crops such as soy beans) and plant molecular marker assisted breeding (i.e. attaining better proprieties such as disease resistance).

Red biotechnology uses the human body's own tools and weapons to fight diseases. Red biotechnology is of great importance in traditional drug discovery and also in creating new possibilities for treatment, prevention and diagnosis (by using new methods). Biotech medicines account for 20% of all market medicines. The continuous growth of knowledge, new discoveries and investments in this field, results in broadening opportunities for curing too. White biotechnology is connected with industrial applications. White biotech uses molds, yeasts, bacteria and enzymes to produce goods, services or products. It offers a wide range of bioproducts like detergents, vitamins and antibiotics. Most of the white biotechnology processes result in the saving of water, energy, chemicals and in the reduction of waste compared to traditional methods. However, this area is not new, since such processes have been used for thousands of years in the production of wine, cheese, bread and many others. Blue biotechnology is the term used to describe aquatic and marine applications of biotechnology (Raju 2016).

Biotechnology applications can be divided into five key sectors: biomedicine, bioagriculture, industrial biotechnology, bioenergy and bioenvironment (Elnashar 2010). A wide range of antibiotics, vitamins, amino acids, fine chemicals and foodstuffs are manufactured using biotechnology (white biotechnology). Detoxification of industrial and domestic waste water is also carried out by biotechnological means. A biotechnological process generally consists of five sections: raw material preparation (biomass), reaction, product recovery, purification and waste disposal (Fig. 1.1).

Biomass is organic matter derived from living, or recently living, organisms and can be derived from different sources such as plants, animals or microorganisms (Fig. 1.2).

Microorganisms are used as biocatalysts to convert biomass into the products of interest. Furthermore, they can act as biobased factories to produce desired chemicals and materials. Different species of fungi, yeast, bacteria and algae are already employed commercially and are frequently the initial choice for the development of

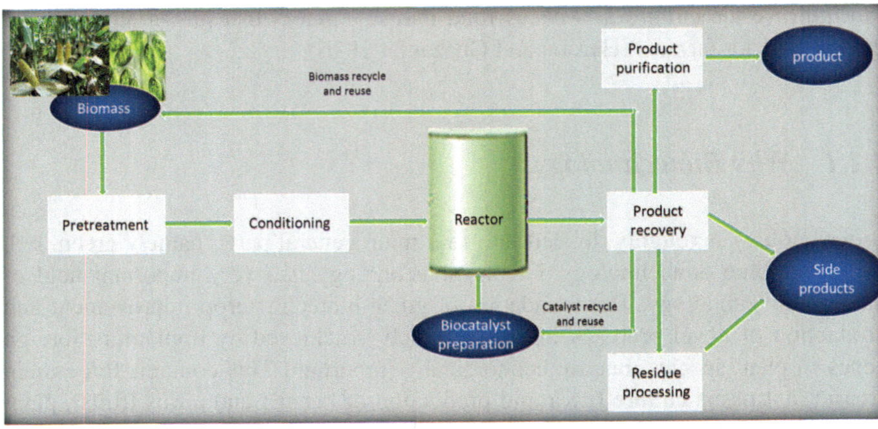

Fig. 1.1 General flow sheet of a biotechnological process (Anarjan et al. 2017)

novel biocatalysts and biobased chemical sources for industrial application. Most active fine chemicals, such as pharmaceuticals, cosmetics, nutritional supplements, flavoring agents as well as additives for foods, feed and fertilizer, are produced enzymatically or through microbial fermentation (Anarjan et al. 2017).

It is well understood that the bioreactor is the heart of any biochemical process, as it dictates both the product quality and the extent of the downstream separation and treatment equipment needed. The bioreactors provide a controlled environment for the production of metabolites which can help to achieve optimum growth of microbes. Bioreactors can be broadly classified into submerge and solid state reactors as shown in Fig. 1.3.

A photobioreactor can also be described as an enclosed, illuminated culture vessel designed for controlled biomass production. Photobioreactor refers to closed systems that are closed to the environment having no direct exchange of gases and contaminants with the environment. Photobioreactors permit the production of complex biopharmaceuticals (Anarjan et al. 2017). Different types of photobioreactors have been designed and developed for the production of algae (Fig. 1.4).

1.2.2 *Food Biotechnology*

Egyptians were brewing beer and producing baked products by the fourth millennium BC. Distillation of ethanol was developed and applied in China in the second millennium. After that, by 5000 and 4000 BC, cheese and vinegar were produced biotechnologically, respectively (Ibrahim and Day 2014).

In fact, for many thousands of years, man has used naturally occurring microorganisms, such as bacteria, yeasts and molds, and their enzymes, to make foods such as bread, cheese, beer and wine. The main applications of biotechnology in food are shown in Fig. 1.5.

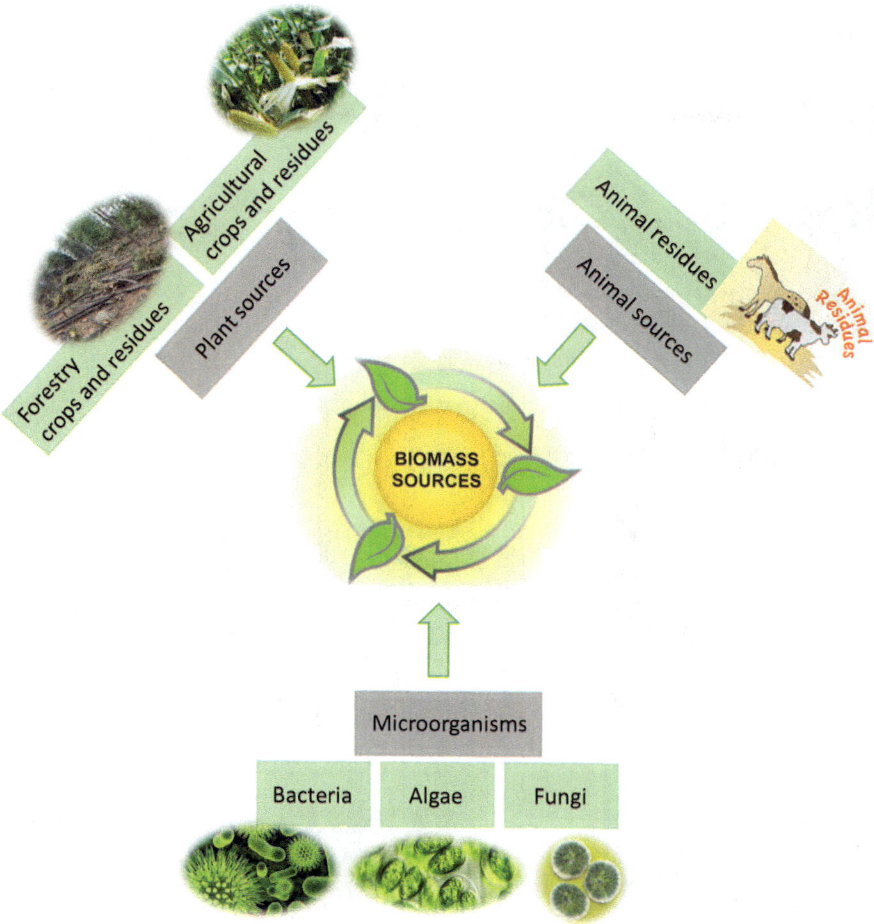

Fig. 1.2 Major sources of biomass (Anarjan et al. 2017)

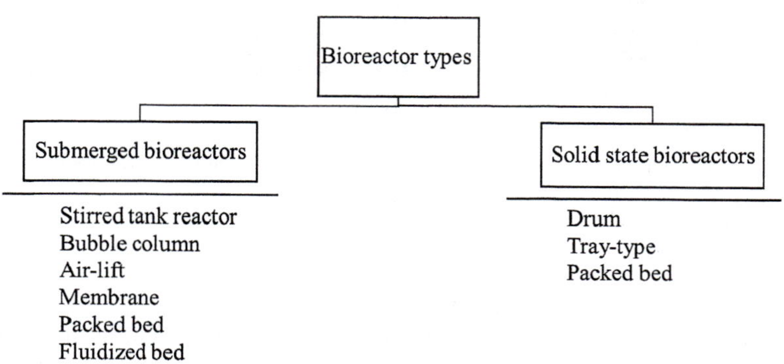

Fig. 1.3 Types of submerged and solid state bioreactors (Anarjan et al. 2017)

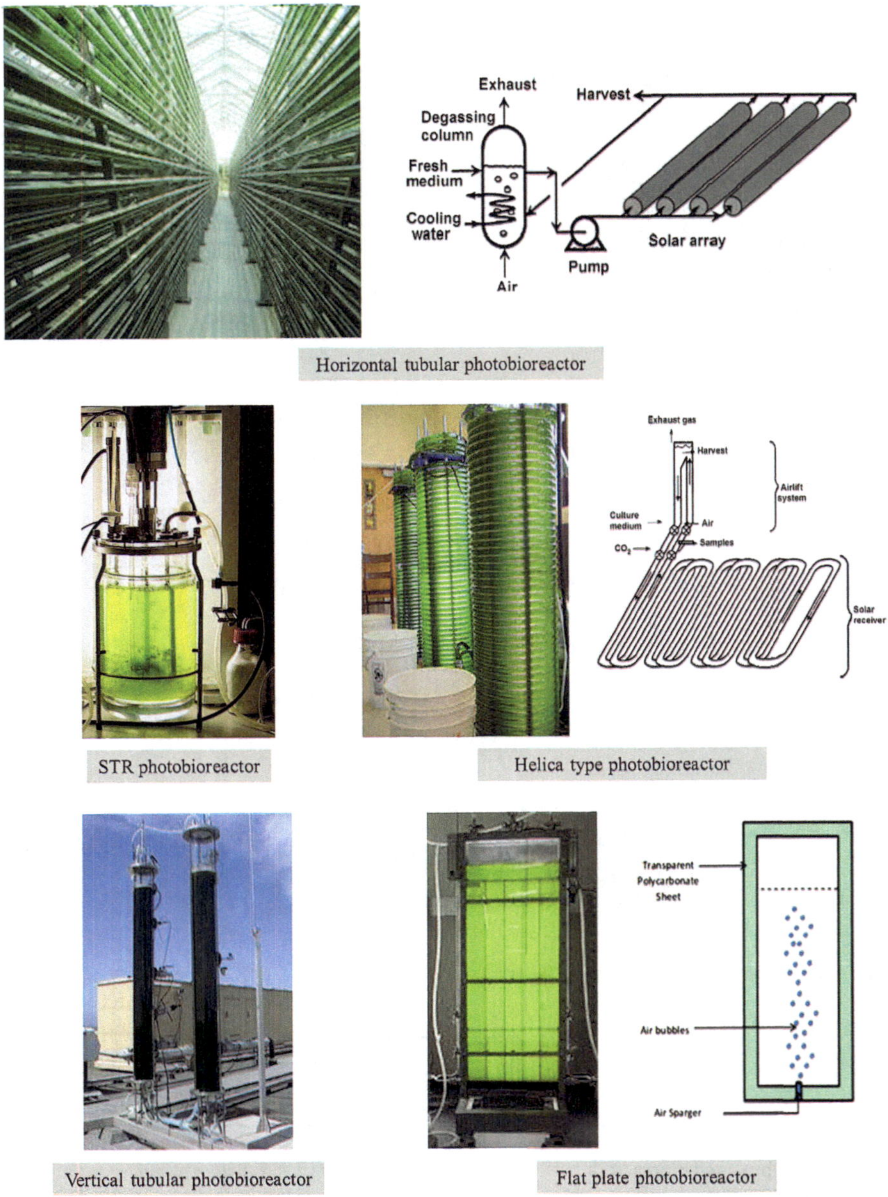

Fig. 1.4 Types of photobioreactors (Anarjan et al. 2017)

Biotechnology as applied to food processing in most developing countries makes use of microbial inoculants to enhance properties such as the taste, aroma, shelf life, texture and nutritional value of foods. The process whereby micro-organisms and their enzymes bring about these desirable changes into food materials is known as

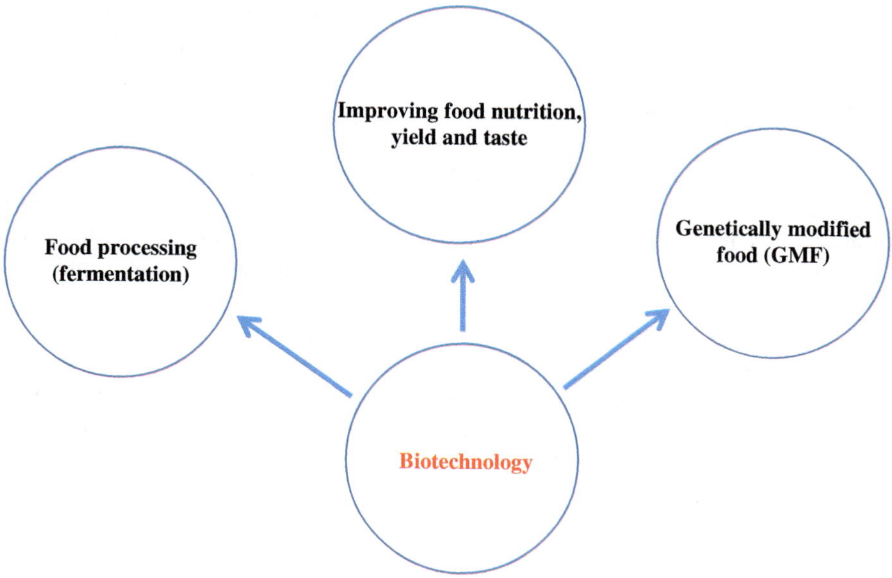

Fig. 1.5 Main applications of biotechnology in food

fermentation. Fermentation processing is also widely applied in the production of microbial cultures, enzymes, flavors, fragrances, food additives and a range of other high value-added products (Haroon and Ghazanfar 2016; Ruane and Sonnino 2011).

Fermentation is a process in which organic compounds act as donor or acceptor of hydrogen, under anaerobic conditions. In the industries, fermentation can be defined as the breakdown or catabolism of organic compounds by microorganisms under both aerobic and anaerobic conditions to produce end products. Genetically modified food is synthesized using biotechnological tools. In the 1980s, recombinant gene technology led to the production of rennet enzyme for cheese production and genetically engineered yeast for baking. These genetically engineered bio ingredients were the first products manufactured using recombinant technology (Ibrahim and Day 2014).

Today genetically engineered microorganisms for the production of vitamins, organic acids, amino acids, sweeteners, edible oils and nutritional supplements can be developed from the insertion of a functional gene (DNA) into a host such as lactic acid bacteria. These bacteria are a Gram-positive bacteria presented in fermented foods and are identified as Generally Recognized as Safe (GRAS). Lactic acid bacteria and probiotic microorganisms in fermented foods have been used for many years for health reasons and are now an attractive alternative for the treating of intestinal disorders and seem to influence the immune system via stimulating protective immune cells. Through genetic engineering, it is possible to strengthen the effect of existing probiotic strains and create completely new probiotics with multiple health benefits (Axelsson et al. 2003).

Biotechnology, including genetic engineering technology, is going to play an important role in the production for functional foods. Functional foods are also known as Nutraceuticals that are going to become preventive medicines to tackle health related issues. For example, it is widely believed that omega-3 fatty acids are beneficial against cardiovascular diseases (Wildman 2002). Breweries are synthesized through the process of fermentation. Yeast strains are used to make breweries at commercial level. Genetic engineering has enabled us to make light wine. Yeast is genetically modified through foreign gene encoding glucoamylase. During the fermentation process, yeast expresses glucoamylase that converts starch into glucose (Lawrence 1988).

Every food item does not contain all essential components. For example, rice is used as a staple food in many countries, but being devoid of vitamin A, it's not a perfect staple food. Use of biotechnological techniques has solved these problems through introduction of the foreign vitamin A gene (Sun 2008). By 2050, the population of the world will be nine billion. Therefore, more yield will be required on the same land. Biotechnology is potentially the best technology to fight against the problem of food yield (Haroon and Ghazanfar 2016; Ruane and Sonnino 2011).

Biotechnology has also allowed scientists to produce fruits with better taste. Genetically modified foods with better taste include seedless watermelon, tomato, eggplant, pepper and cherries. Elimination of seeds from these food articles resulted in more soluble sugar content, enhancing sweetness. Fermentation pathways are modified using biotechnology to add aroma in wine (Falk et al. 2002). Today, enzymes are used for an increasing range of applications in bakery, cheese-making, starch processing and production of fruit juices and other drinks. They can improve texture, appearance and nutritional value, and may generate desirable flavors and aromas. Currently-used food enzymes sometimes originate in animals or plants (for example, a starch-digesting enzyme, amylase, can be obtained from germinating barley seeds), but most come from a range of beneficial microorganisms. In the bread-making process, amylase is used to break down flour into soluble sugars, which are transformed by yeast into alcohol and carbon dioxide. This makes the bread rise (Shoseyov and Levy 2008). Table 1.1 indicates some of the main enzymes that are used in food industries. Some of the indigenous fermented foods in Southeast Asia are shown in Table 1.2.

1.3 Nanotechnology Definition

The term 'nano' is derived from the Greek word for dwarf. The term 'nanotechnology' was first used in 1974 by the late Norio Taniguchi and concepts were given by Richard Feynman in 1959 (Sundarraj et al. 2014). Nano is a prefix that means 'one-billionth'. The nanometer is one-billionth of a meter—much too small to see with the naked eye or even with a conventional light microscope. Nanotechnology involves creating and manipulating materials at the nano scale (Hansen et al. 2013).

Table 1.1 Some of the food enzymes and their applications (Shoseyov and Levy 2008)

Enzymes	Food applications
Rennet	Coagulant in cheese production
Lactase	Hydrolysis of lactose to give lactose-free milk products
Protease	Hydrolysis of whey proteins—Breakdown of proteins—Meat tenderizing
Catalase	Removal of hydrogen peroxide in dairy products
Cellulases, beta-glucanases, alpha amylases, proteases, maltogenic amylases	For liquefaction, clarification and to supplement malt enzymes—Breakdown of starch and maltose production—Delays process by which bread becomes stale—Production of glucose syrups
Amyloglucosidase	Conversion of starch to sugar for alcohol production and saccharification
Pentosanase	Breakdown of pentosan, leading to reduced gluten production
Glucose oxidase	Stability of dough—Oxygen removal from juice
Pectinase	Increase of yield and juice clarification
Inulinase	Production of fructose syrups

Richard Feynman predicted the emergence of a new science called nanotechnology, a branch of science that deals with structures of 1–100 nm in scale. According to the National Nanotechnology Initiative, nanotechnology is the understanding and control of matter at dimensions of roughly 1–100 nm, where unique phenomena enable novel applications (Ravichandran 2009).

One nanometer is about 60,000 times smaller than a human hair in diameter or the size of a virus. A typical sheet of paper is about 100,000 nm thick, a red blood cell is about 2000–5000 nm in size, and the diameter of DNA is in the range of 2.5 nm. Therefore, nanotechnology deals with matter that ranges from one half the diameter of DNA up to 1/20 the size of a red blood cell (Sundarraj et al. 2014).

1.3.1 Why Nanotechnology?

Nanotechnology means manipulation of material at a very small scale, usually less than 100 nm. This miniaturization leads to very impressive properties and functions. Nanomaterials can be found in many forms: nanoparticles, nano rods, nano tubes, nano sheets, nanofibers, etc. Nowadays, nanotechnology has been applied in many sciences and technology such as electronics, energy, catalysts, agriculture and the food industry (Assa et al. 2015).

Reduction of particle sizes into nano scale changes surface area, solubility, delivery properties, absorption by cells and the residence time in the body. Some of these properties, such as high surface area, increase the efficiency of biomaterials to reduce the risk of certain diseases such as cancer (Assa et al. 2017).

Table 1.2 Indigenous fermented foods of Southeast Asia (Owens 2014)

Product	Geography	Substrate	Microorganisms	Product use
Ang-kak	Indonesia	Rice	*Monascus purpureus*	Colorant
Bagoong	Philippines	Fish	Bacteria	Seasoning agent
Bonkrek	Indonesia	Coconut press cake	*Rhizopus oligosporus*	Meat substitute
Fish sauce	Southeast Asia	Fish	Bacteria	Seasoning agent
Kecap	Indonesia	Soybean, wheat	*Aspergillus oryzae, Lactobacillus, Hansenula, Saccharom*	Condiment and seasoning agent
Lao-chao	Indonesia	Rice	*Rhizopus oryzae, R. chinensis*	Eaten as dessert or combined with seafood
Monosodium glutamate	Malaysia	Starch, sugar	*Brevibacterium glutamicum*	Seasoning agent
Nata de coco	Malaysia Philippines Indonesia	Coconut milk	*Acetobacter xylinum*	Dessert
Oncom	Indonesia	Peanut press cake	*Neurospora intermedia, Rhizopus oligosporus*	Roasted or fried in oil, used as meat substitute
Peujeum	Indonesia	Banana, plantain	Bacteria	Eaten fresh or fried
Puto	Philippines	Rice	*Lactic acid bacteria, Saccharomyces cerevisiae*	Snack
Sapal	Papua New Guinea	Taro corm, coconut cream	*Leuconostoc mesenteroides, Lc. paramesenteroides*	Seasoning agent
Soy sauce	Malaysia Philippines Indonesia Thailand	Soybeans and wheat	*Aspergillus oryzae, A. soyae, Lactobacillus bacteria, Zygosaccharomyces rouxii*	Seasoning for meat, fish, cereals, vegetables
Tao-si	Philippines	Soybeans and wheat flour	*Aspergillus oryzae*	Seasoning agent
Tauco	Indonesia	Soybeans, cereals	*Rhizopus oligosporus, Aspergillus oryzae*	Drink
Tempe	Indonesia Malaysia	Soybeans	*Rhizopus oligosporus*	Fried in oil, roasted, meat substitute in soup
Tofu	Malaysia Indonesia	Soy milk	*Monascus purpureus*	Seasoning agent

1.3.2 Food Nanotechnology

Nanotechnology is the science of manufacturing and application of materials and structures which are at the scale of nanometer. In recent years, significant investments in the field of nanotechnology have been made by food and agricultural industries. These efforts have improved the quality of products and have reduced the total costs (Assa et al. 2015).

In food engineering, two major applications related to nanotechnology are food nano sensing and food nanostructured ingredients. In the former field, better food quality and safety evaluation can be achieved by using nanotechnology. Recently, nanotechnology is finding its way into dairy and food processing, preservation, packaging and functional foods development (Sundarraj et al. 2014).

Organic materials, which naturally exist in foods such as carbohydrate, fat, vitamins and proteins, can be in different sizes such as large macromolecules or a simple mono molecule in the nano range. Due to special properties of biological nano materials, they can be used for various purposes to improve taste, texture and sustainability (Assa et al. 2015).

There are two known categories for application of nanotechnology, especially in food industries. These include 'top down' and 'bottom up' systems. The top down approach means breaking down of large particles to materials that are in range of nanometric dimensions. This method of manufacturing of nanoparticles is basically included in the physical and mechanical processing such as grinding and milling (Sanguansri and Augustin 2006). For example, dry milling of wheat to flour increases the water binding capacity of flour (Degant and Schwechten 2002; Zhu et al. 2010). Moreover, size reduction technology has been used to improve the antioxidant effect of green tea. When the particle size of green tea is decreased to about 1000 nm, the high ratio of nutrient digestion and absorption leads to high activity of the oxygen eliminating enzyme (Shibata 2002). The homogenization process, which is widely used in diary industries, is another example of the top down size reduction mechanism. In this process the size of fat globules are reduced by applying pressure (Bud 1989).

On the other hand, the self-organization and self-assembly of biological compound is categorized in a bottom up approach. Crystallization, layer by layer deposition, microbial synthesis, biomass reaction and solvent extraction-evaporation are the methods which can be used in a bottom up way of manufacturing of nanomaterials. In this method, molecules arrange step by step with specific features. For example, the casein micelles can result in a stable nanomaterial by their self-assembly. A balance between various non-covalent forces can lead to self-organized biological entities on the nanometer scale (Sozer and Kokini 2009).

Bioactive proteins which are used in functional food for their health benefits are one of these applications. Their reduced size helps them to improve their availability and solubility and as a result their ability to be transferred across intestinal membranes (Shegokar and Müller 2010). In addition, nanoemulsions, which have

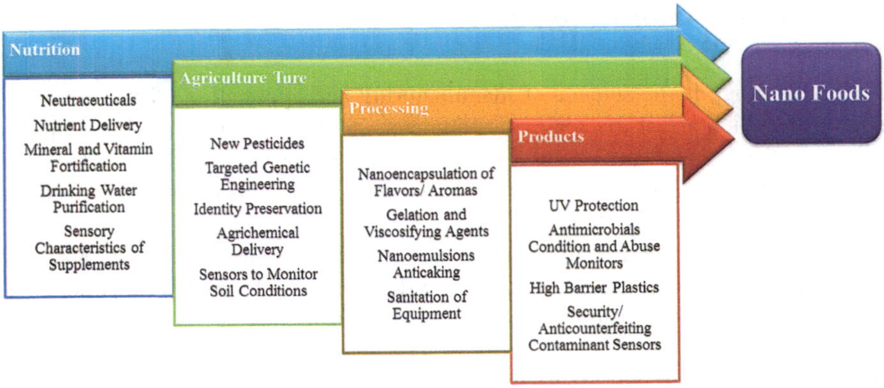

Fig. 1.6 Nanotechnology rule in the food industry (Assa et al. 2015)

significant rheological and textural properties, can be used in food products to reduce their fat content with no change in their creaminess. Figure 1.6 shows the main application of nanotechnology in the food industry.

1.4 Nanobiotechnology Definition

Nanobiotechnology is an emerging field of research at the crossroads of biology and nanoscience. It is involved in many different disciplines, including physicists, chemists, engineers, information technologies and material scientists as well as biologists. Nanobiotechnology incorporates biotechnology on the nano-scale size. This research field includes two approaches. One is the application of the tools and processes of nanotechnology to study and manipulate biological systems, and the other is the use of biological systems as templates in the development of nano-scale products. In fact, nanobiotechnology is the intersection of inorganic and organic engineering to solve critical problems in biology (Niemeyer and Mirkin 2004).

1.4.1 Why Nanobiotechnology?

The field of nanobiotechnology is growing day by day in regard to drug delivery, cosmetics and environmental applications. The surface to volume ratio of the nano-sized particles is high compared to micro and macro sized particles. Hence, they are easily attracted to the biological environment and delivered the target particles to target site (Satyavani and Gurudeeban 2014).

Biosynthesis of nanoparticles of environmentally benign materials, like plant, microalgae, bacteria, fungi and animals, has been increased. Green synthesis provides advancement over the chemical and physical method as it is cost effective, environmentally friendly, easy to scale up and there is no need to use high pressure, energy, temperature and toxic chemicals (Mohammadlou et al. 2016).

Biodegradable nanoparticles are frequently used to improve the therapeutic value of various water-soluble and insoluble medicinal drugs and bioactive molecules by improving bioavailability, solubility and retention time. These nanoparticle drug formulations reduce patient expenses and the risk of toxicity. An exciting potential solution in cancer treatments is to encapsulate the drug in a biocompatible material that can be injected into the blood stream with the intention of delivering the drug to a tumor site. Polysaccharides, lipids, surfactants and dendrimers have received increasing attention due to their outstanding physical and biological properties (Ghaz-Jahanian et al. 2015). Figure 1.7 indicates several nanocarriers used in targeted drug delivery. Iron oxide nanoparticles have been used widely in various medicinal areas. Figure 1.8 shows the main applications of these nanoparticles in medicine fields. Nanosensors are emerging as promising tools for applications in the agriculture and food production. They offer significant improvements in selectivity, speed and sensitivity compared to traditional chemical and biological methods. Nanosensors can be used for determination of microbes, contaminants, pollutants and food freshness. The nanosensors used in food analysis combine knowledge of biology, chemistry and nanotechnology and may also be called nanobiosensors (Omanović-Mikličanina and Maksimović 2016).

1.4.2 Food and Nanobiotechnology

The research in the area of nanobiotechnology in food involves mainly adding antioxidants, antimicrobial, biosensors and other nanomaterials in food materials. Medical, pharmaceutical and cosmetics industries use nanoparticles made from food to improve the characteristics of the products. Nanobiotechnology in food packaging has been a focus in recent years. The potential perspectives of bionanocomposites for food packaging applications together with bio-based materials, such as edible and biodegradable nanocomposite films, have gained significant attention. Among the available metal nanoparticles, silver and related materials have been utilized in many nano-based commercial products for their antimicrobial property. Studies suggest that the antimicrobial performance is enhanced due to an intensive surface area/reduced particle size.

Biosurfactants are surface active substances that can reduce interfacial tension and are produced or excreted at the microbial cell surface. Biosurfactants have been tested in environmental applications, cosmetics, foods and pharmaceutical industries but also as industrial cleaners and chemical products for agricultural use. The main components of the food micro and nanoemulsions are oil, water and surfactant. The surfactant is used to create a low interfacial tension, which aids in the production of nanosized particles. There is a need to monitor food-borne pathogens throughout the food chain from production, processing and distribution to the point-of-sale. Pathogens may be present in low numbers in a sample for analysis that

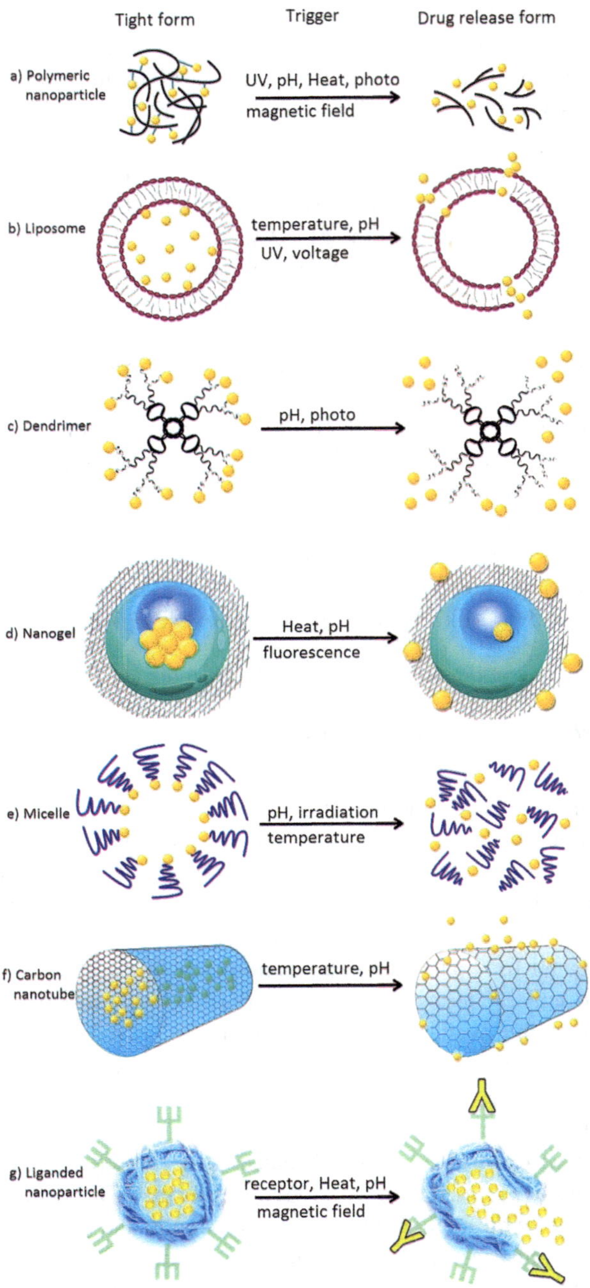

Fig. 1.7 Several nanocarriers in targeted drug delivery (Ghaz-Jahanian et al. 2015)

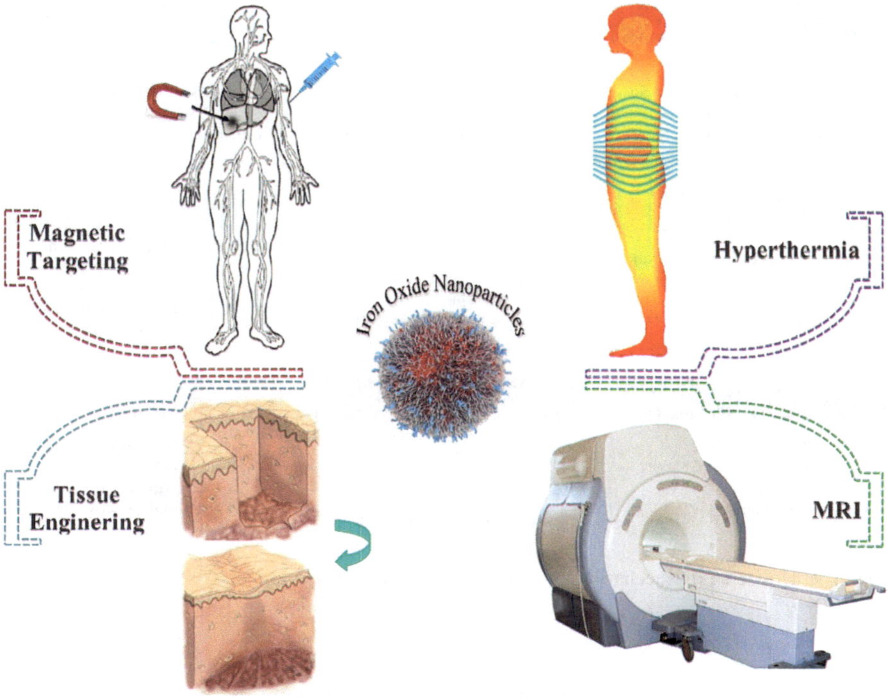

Fig. 1.8 Applications of iron oxide nanoparticles in medicinal field (Assa et al. 2017)

makes the detection difficult. Traditional detection methods for pathogen determination, like colony count estimation, can be laborious and time consuming with completion ranging from 24 h for *E. coli* to 7 days for *Listeria monocytogenes*. These pose significant difficulties for quality control of semi-perishable foods. Advances in the manipulation of nanomaterials permit binding of different biomolecules such as bacteria, toxins, proteins and nucleic acids. One of the major advantages of using nanomaterials for biosensing is that due to their large surface area, a greater number of biomolecules are allowed to be immobilized, which consequently increases the number of reaction sites available for interaction with a target species. This property, coupled with excellent electronic and optical properties, facilitates the use of nanomaterials in label-free detection and in the development of biosensors with enhanced sensitivities and improved response times (de Morais et al. 2014).

Magnetic nanoparticles, especially iron magnetic nanoparticles, can be used in food analysis, protein/enzyme immobilization, protein purification and water treatment. These types of nanoparticles have hydrophobic surfaces and large surface area to volume ratio which tend to agglomerate in both biological medium and magnetic field and create heterogeneous size distribution patterns. This limits their

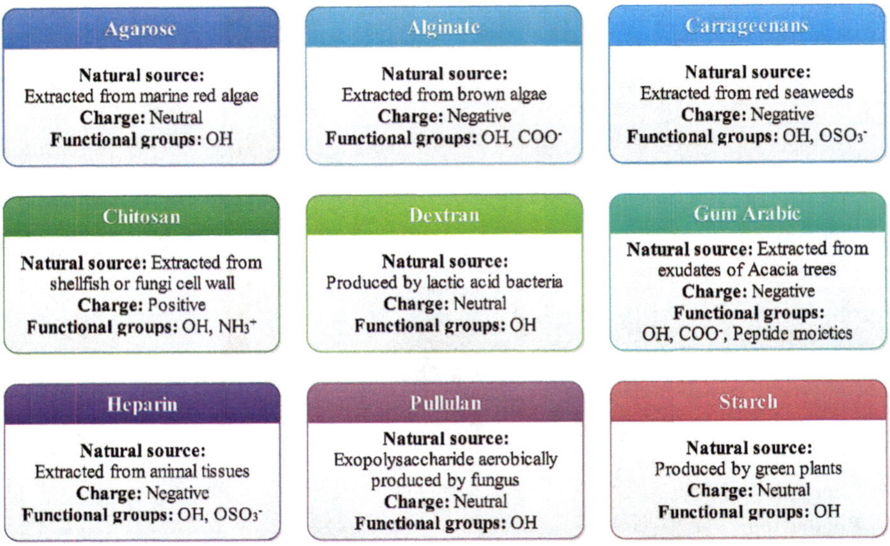

Fig. 1.9 Utilized polysaccharides in magnetic nanoparticles coating and encapsulation (Assa et al. 2016)

applications in different areas such as medical applications. By coating or encapsulation of the magnetic nanoparticles, it is possible to overcome the mentioned problems (Assa et al. 2016). Figure 1.9 shows the polysaccharides which are commonly used for coating and encapsulation of the magnetic nanoparticles.

References

Anarjan N, Vaghari H, Jafarizadeh-Malmiri H, Berenjian A. Intensification of bio-based processes—bioreactors models. Advances in energy research, Volume 28. New York: Nova Science Publishers; 2018. p. 111–46.

Assa F, Jafarizadeh-Malmiri H, Anarjan N, Berenjian A, Ghasemi Y. Applications of chitosan nanoparticles in active biodegradable and sustainable food packaging. Renewable energy and sustainable development. New York: Nova Science Publishers, Inc; 2015. p. 227–44.

Assa F, Jafarizadeh-Malmiri H, Ajamein H, Anarjan N, Vaghari H, Sayyar Z, Berenjian A. A biotechnological perspective on the application of iron oxide nanoparticles. Nano Res. 2016;9(8):2203–25.

Assa F, Jafarizadeh-Malmiri H, Ajamein H, Vaghari H, Anarjan N, Ahmadi O, Berenjian A. Chitosan magnetic nanoparticles for drug delivery systems. Crit Rev Biotechnol. 2017;37(4):492–509.

Axelsson L, Lindstad G, Naterstad K. Development of an inducible gene expression system for Lactobacillus sakei. Lett Appl Microbiol. 2003;37(2):115–20.

Bud R. Janus-faced biotechnology:an historical perspective. Trends Biotechnol. 1989;7(9):230–3.

de Morais MG, Martins VG, Steffens D, Pranke P, da Costa JAV. Biological applications of nanobiotechnology. J Nanosci Nanotechnol. 2014;14(1):1007–17.

Degant O, Schwechten D. Wheat flour with increased water binding capacity and process and equipment for its manufacture. German Patent DE10107885A1; 2002.

Elnashar MM. Immobilized molecules using biomaterials and nanobiotechnology. J Biomater Nanobiotechnol. 2010;1(1):61–77.
Falk MC, Chassy BM, Harlander SK, Hoban TJ IV, McGloughlin MN, Akhlaghi AR. Food biotechnology: benefits and concerns. J Nutr. 2002;132(6):1384–90.
Ghaz-Jahanian MA, Abbaspour-Aghdam F, Anarjan N, Berenjian A, Jafarizadeh-Malmiri H. Application of chitosan-based nanocarriers in tumor-targeted drug delivery. Mol Biotechnol. 2015;57(3):201–18.
Govil C, Aggarwal A, Sharma J. Plant biotechnology and genetic engineering. New Delhi: PHI Learning Pvt. Ltd; 2017.
Hansen SF, Howard CV, Martuzzi M, Depledge M. Nanotechnology and human health: scientific evidence and risk governance: report of the WHO expert meeting 10–11 December 2012, Bonn, Germany; 2013.
Haroon F, Ghazanfar M. Applications of food biotechnology. J Ecosys Ecograph. 2016;6(215):2.
Ibrahim O, Day D. Biotechnology in nutrition and food engineering. J Nutr Health Food Eng. 2014;1(5):0026.
Lawrence RH. New applications of biotechnology in the food industry. In: National Research Council (US) Commission on Life Sciences, editor. Biotechnology and the Food Supply Proceedings of a Symposium. Washington, DC: National Academies Press; 1988. p. 19–45.
Mohammadlou M, Maghsoudi H, Jafarizadeh-Malmiri H. A review on green silver nanoparticles based on plants: synthesis, potential applications and eco-friendly approach. Int Food Res J. 2016;23(2):446–63.
Niemeyer CM, Mirkin CA. Nanobiotechnology: concepts, applications and perspectives, vol. 1. New York: Wiley; 2004.
Omanović-Mikličanina E, Maksimović M. Nanosensors applications in agriculture and food industry. Bull Chem Technol Bosnia Herzegovina. 2016;47:59–70.
Owens JD. Indigenous fermented foods of Southeast Asia. Boca Raton: CRC Press; 2014.
Raju P. World history of modern biotechnology and its applications. Biotechnol Ind J. 2016;12(11):107–13.
Ravichandran R. Nanoparticles in drug delivery: potential green nanobiomedicine applications. Int J Green Nanotechnol Biomed. 2009;1(2):B108–30.
Ruane J, Sonnino A. Agricultural biotechnologies in developing countries and their possible contribution to food security. J Biotechnol. 2011;156(4):356–63.
Sanguansri P, Augustin MA. Nanoscale materials development—a food industry perspective. Trends Food Sci Technol. 2006;17(10):547–56.
Satyavani K, Gurudeeban S. Green revolution towards nanobiotechnology. J Nanomed Res. 2014;2(1):14.
Shegokar R, Müller RH. Nanocrystals: industrially feasible multifunctional formulation technology for poorly soluble actives. Int J Pharm. 2010;399(1–2):129–39.
Shibata T. Method for producing green tea in microfine powder. United States Patent US6416803B1. 2002.
Shoseyov O, Levy I. Nanobiotechnology: bioinspired devices and materials of the future. Berlin: Springer; 2008.
Sozer N, Kokini JL. Nanotechnology and its applications in the food sector. Trends Biotechnol. 2009;27(2):82–9.
Sun SS. Application of agricultural biotechnology to improve food nutrition and healthcare products. Asia Pac J Clin Nutr. 2008;17(S1):87–90.
Sundarraj A, GuhanNath S, Aaron S, Ranganathan T. Recent innovations in nanotechnology in food processing and its various applications—a review. Int J Pharm Sci Rev Res. 2014;29(2):116–24.
Wildman RE. Handbook of nutraceuticals and functional foods. Boca Raton: CRC Press; 2002.
Zhu K, Huang S, Peng W, Qian H, Zhou H. Effect of ultrafine grinding on hydration and antioxidant properties of wheat bran dietary fiber. Food Res Int. 2010;43(4):943–8.

Chapter 2
Challenges for Nanobiotechnology

2.1 Introduction

Nanomaterials are making their way into all aspects of our lives; these materials are being increasingly used in pharmaceutical and medical applications, cosmetics and personal products, energy storage and efficiency, water treatment and air filtration, environmental remediation, chemical and biological sensors, military defense and explosives, and in countless consumer products and materials. For instance, in the area of food, nanomaterials can be used to provide new tastes and flavors, functional foods, hygienic food processing and packaging, intelligent, lightweight and strong packaging systems, pathogen detection and removal, and reduced agrochemicals, colors, flavors and preservatives (World Health Organization, Regional Office for Europe 2013).

A number of reviews have recognized the vast opportunities for applying nanotechnology to agriculture and to all aspects of the food industry, providing preservation, processing, packaging and monitoring functions (Fig. 2.1) (Handford et al. 2014).

Nanobiotechnology is a recently coined term describing the convergence between engineering and molecular biology. Nanobiotechnology has gained much attention these days due to its wide applications in the food industry. It has applications in enzyme immobilization, green synthesis of inorganic nanoparticles, preparation of nanoemulsions and encapsulation, nanosensors, and packaging (de Morais et al. 2014).

The extended shelf life of food products is also possible through innovative packaging that incorporates antimicrobial properties. This application offers huge potential to the food industry by contributing to a reduction in food waste, as well as a better quality and safer food supply. In addition, the use of nanosensors in food packaging for detection of food spoilage is important to combating pathogenic microorganisms and, consequently, reducing foodborne illnesses in consumers (Handford et al. 2014).

Despite a rapid development of nanomaterials and their uses in the food sector, little is known about their *in vivo* and *in vitro* kinetics. Consequently, the risks of

H. Jafarizadeh-Malmiri et al., *Nanobiotechnology in Food: Concepts, Applications and Perspectives*, https://doi.org/10.1007/978-3-030-05846-3_2

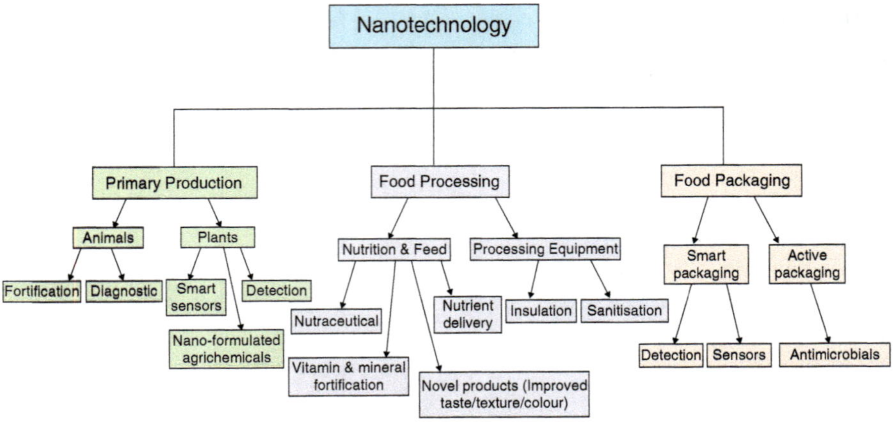

Fig. 2.1 Main applications and opportunities of nanotechnology in agri-food areas (Handford et al. 2014)

nanomaterials have not been analyzed. Nanoparticles are reported to be absorbed across the intestinal barrier via transcellular, paracellular, and junctional pathways, but the bioavailability of each material may be different due to various factors. Questions about these compounds have already raised safety concerns, although the history of their use in the food sector is yet short (Higashisaka et al. 2015).

This chapter overviews the information currently available about nanomaterials' safety and toxicity aspects, marketing concerns, risk assessments and regulatory aspects.

2.2 Safety Aspects

The major nanomaterials used in consumer products are silver, silica, iron and titanium dioxide. Among different product categories, silver nanoparticles (Ag NPs) are the most widely used and particularly common in food and beverage products (Mohammadlou et al. 2016). Owing to the antimicrobial activity of silver, Ag NPs are widely used in food products. The mechanism of the antibacterial activity of Ag NPs has not yet been elucidated, but they may interact with the membranes of bacteria. The antibacterial activity of Ag NPs is likely due to the formation of Ag ions on the surface of the NPs through the reaction with oxygen. The antibacterial activity of Ag NPs increases with decreasing particle size, which has been attributed to the increase in the surface area to mass ratio as particle size decreases (Higashisaka et al. 2015).

Although useful and exciting, the incorporation of Ag NPs in food related applications is topical to the concerns surrounding food nanotech risk perceptions, in that some issues have arisen. Several reports have indicated that Ag NPs are toxic to cells, and can alter the normal function of mitochondria, increase membrane permeability and generate reactive oxygen species (AshaRani et al. 2008; Bryksa and Yada 2012).

Amorphous silica NPs are widely used in food products, for example, as thickeners, anti-caking agents and carriers for fragrances and flavors. Nanosilica (E551) is registered as a food additive in the European Union; Dekkers et al. (2011) analyzed food products, including E551, and found that they contained silica nanoparticles at concentrations ranging from <0.1 to 6.9 mg/g product and with particle sizes ranging from 30 to 200 nm (Higashisaka et al. 2015).

Titanium dioxide NPs are one of the top five types of nanomaterials used in consumer products such as cosmetics, as well as in food products, paints and medicines. Titanium dioxide has been approved by USFDA and is widely used as a white pigment and food colorant. The daily intake of titanium dioxide from food is estimated to be 5.4 mg/person/day (about 90 μg/kg bw/day) in the United Kingdom. In Europe, food-grade titanium dioxide is designated as E171, and approximately 36% of the particles in E171 have diameters of <100 nm. Candies, sweets and chewing gums have the highest content of titanium (0.01–1 mg titanium per serving). Several studies have indicated non-toxicity of TiO_2 NPs against human and animal cells (Can et al. 2011; Higashisaka et al. 2015).

Magnetic iron oxide nanoparticles (IONPs) have potential applications in various disciplines of science ranging from environmental remediation to biomedical such as magnetic drug targeting, tissue repair, cell tissue targeting and enzyme immobilization. Super paramagnetic iron oxide nanoparticles (SPIONs) are small synthetic α-Fe_2O_3, γ-Fe_2O_3 or Fe_3O_4 particles with a core diameter ranging from 10 to 20 nm. SPIONs have minimal toxicity in the human body (Assa et al. 2016).

A study comparing several metal oxide of NPs showed IONPs to be safe and non-cytotoxic at concentrations below 100 mg/ml. However, intravenous administration can lead to accumulation in a targeted organ, potentially leading to iron overload, which can be toxic. High free iron levels can cause an imbalance in homeostasis, leading to DNA damage, oxidative stress and inflammation. Based on these reports, we believe that these particles can be used safely in humans provided that concentrations are maintained below 100 mg/ml and accumulation in organs is monitored to prevent iron overload. Special attention should also be given to leaching of iron (Fe^{+3}) ions and interaction of these particles with H_2O_2, which could generate free radicals such as hydroxy radicals due to Fenton chemistry (Buyukhatipoglu and Clyne 2011). Figure 2.2 shows SPIONs toxicity.

2.3 Public Acceptance

Consumer acceptance of new food technologies is influenced by many factors, including consumers' perceptions of the risks and benefits as well as perceived quality, perceived naturalness, price, and general attitudes, values and cultural norms. Consumer acceptance of nanotechnology is clearly a key marketing concern. Consumers' trust in industry and government as a source of nanotechnology

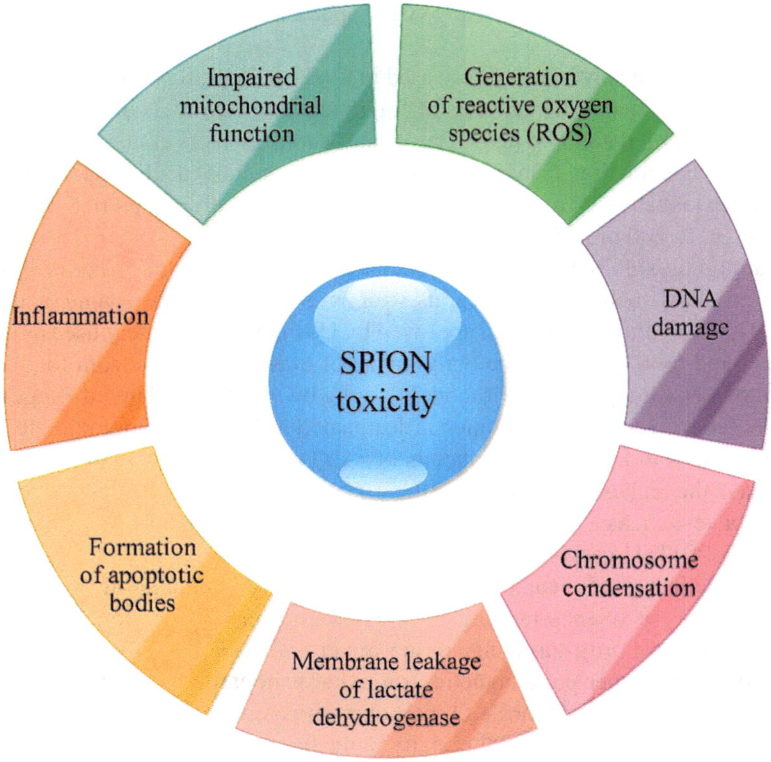

Fig. 2.2 SPIONs toxicity (Buyukhatipoglu and Clyne 2011)

information and oversight will be important when they frame their impressions on perceived risks and benefits of nanotechnology for food applications. Recent literature suggests that there is low public trust in both industry and government when it comes to nanotechnology (Buzby 2010).

Increasing scientific evidence has demonstrated that exposure to NPs (e.g., silver, silicates, titanium dioxide and iron oxide) may lead to oxidative damage and inflammatory reactions of the GI tract. Furthermore, long-term exposure to NPs has been linked to acute toxic response, including lesions of the kidney and liver, as well as numerous forms of cancer. Several reviews have reported uncertainties regarding the potential adverse effects of nano-packaging materials on human health. Discussion in these studies indicates that the main risk of consumer exposure to NPs from food packaging materials is indirectly through the possible migration into foodstuffs or ingestion of edible coatings (Chaudhry and Castle 2011; Han et al. 2011; Kuzma et al. 2008).

NPs can also enter the central nervous system either directly through axons of the olfactory pathway or through systemic circulation through the olfactory bulb. An important part of consumer acceptance of agri-food nanotechnology is societal inclusivity in the process of product design, development and commercialization of different applications. There are many ways to collate information about societal preferences and priorities, for example through qualitative and quantitative research which can be applied to 'fine-tune' the final delivery of different applications to the consumer (Frewer et al. 2014).

2.4 Risk Assessment

The five main challenges are to develop instruments to assess exposure to engineered nanomaterials. It is fairly understood that exposure of humans and animals to the environment potentially contaminated with nanomaterials may need to be monitored for any adverse consequence. The challenge becomes increasingly difficult in more complex matrices like food. The second challenge would be to develop applicable methods to detect and determine the toxicity of engineered nanomaterials within the next 5–15 years. Then again, proposing models for predicting effects of these nanomaterials on human health and the environment would be an inevitable issue. The next challenge would be to develop reverse systems to evaluate the precise impact of engineered nanomaterials on health and the environment over the entire life span that speaks to the life cycle issue. The fifth being more of a grand challenge would be to develop the tools to properly assess risk to human health and to the environment (Fakruddin et al. 2012).

There are multiple possible primary and secondary exposure pathways stemming from current and potential nanotechnology applications, leading to occupational and consumer exposure. This exposure can occur via inhalation, ingestion or skin absorption depending on the nanomaterial and the specific application (for treated patients, injection is also relevant) (Hansen et al. 2008; Poland 2012). Therefore, there is an urgent need to assess the level of population exposure to nanomaterials, over time and for different population subgroups. Normally, exposure assessment would involve an estimation of the concentrations by mass to which workers, consumers and other environmental receptors are exposed, through all different pathways. Extensive data is required to complete a full exposure assessment including information about manufacturing conditions, level of production, industrial applications and uses, consumer products and behavior, and environmental fate and distribution. Unfortunately, such detailed information is lacking for virtually every type of nanomaterial or group of nanomaterials, and technical difficulties hamper accurate measurement of nanomaterials in the workplace as well as in the environment (World Health Organization, Regional Office for Europe 2013).

Risk assessment guidelines for the use of nanomaterials in the food sector were announced by EFSA in 2011 and by USFDA in 2012. These agencies stressed the importance not only of hazard analyses but also of analysis of the associations between intestinal absorption and characteristics such as particle size and surface properties. Since then, little progress has been made in terms of safety assessments of nanomaterials in the food sector worldwide (Higashisaka et al. 2015).

2.5 Regulatory Aspects

The success of the advancements of nanotechnology in the agri-food industry depends on the consideration of regulatory issues. Legislation is essential to manage potential adverse effects, mitigate risks and to protect consumers. Various government agencies worldwide are becoming increasingly interested in the use of nanotechnology in the food sector. Problems arising from nanofood applications are shown in the practically non-existent laws to regulate this use. There are currently no international regulations of nanotechnologies or nanoproducts (Handford et al. 2014). In short, existing legislation for nanomaterials are inadequate due to existing uncertainties emerging from the difficulty to detect, measure and characterize nanomaterials in food, which are severely hindering risk assessment and exposure assessment. Nevertheless, regulatory considerations will ultimately dictate the success of nanotechnology in food applications.

References

AshaRani P, Low Kah Mun G, Hande MP, Valiyaveettil S. Cytotoxicity and genotoxicity of silver nanoparticles in human cells. ACS Nano. 2008;3(2):279–90.

Assa F, Jafarizadeh-Malmiri H, Ajamein H, Anarjan N, Vaghari H, Sayyar Z, Berenjian A. A biotechnological perspective on the application of iron oxide nanoparticles. Nano Res. 2016;9(8):2203–25.

Bryksa BC, Yada RY. Challenges in food nanoscale science and technology. J Food Drug Anal. 2012;20:418–21.

Buyukhatipoglu K, Clyne AM. Superparamagnetic iron oxide nanoparticles change endothelial cell morphology and mechanics via reactive oxygen species formation. J Biomed Mater Res Part A. 2011;96(1):186–95.

Buzby JC. Nanotechnology for food applications: more questions than answers. J Consumer Affairs. 2010;44(3):528–45.

Can E, Kizak V, Kayim M, Can SS, Kutlu B, Ates M, Kocabas M, Demirtas N. Nanotechnological applications in aquaculture-seafood industries and adverse effects of nanoparticles on environment. J Mater Sci Eng. 2011;5(5):605–9.

Chaudhry Q, Castle L. Food applications of nanotechnologies: an overview of opportunities and challenges for developing countries. Trends Food Sci Technol. 2011;22(11):595–603.

de Morais MG, Martins VG, Steffens D, Pranke P, da Costa JAV. Biological applications of nanobiotechnology. J Nanosci Nanotechnol. 2014;14(1):1007–17.

Dekkers S, Krystek P, Peters RJ, Lankveld DP, Bokkers BG, van Hoeven-Arentzen PH, Bouwmeester H, Oomen AG. Presence and risks of nanosilica in food products. Nanotoxicology. 2011;5(3):393–405.

Fakruddin M, Hossain Z, Afroz H. Prospects and applications of nanobiotechnology: a medical perspective. J Nanobiotechnol. 2012;10(1):31.

Frewer L, Gupta N, George S, Fischer A, Giles E, Coles D. Consumer attitudes towards nanotechnologies applied to food production. Trends Food Sci Technol. 2014;40(2):211–25.

Han W, Yu Y, Li N, Wang L. Application and safety assessment for nano-composite materials in food packaging. Chin Sci Bull. 2011;56(12):1216–25.

Handford CE, Dean M, Spence M, Henchion M, Elliott CT, Campbell K. Nanotechnology in the Agri-Food industry on the island of Ireland: applications, opportunities and challenges; 2014.

Hansen SF, Michelson ES, Kamper A, Borling P, Stuer-Lauridsen F, Baun A. Categorization framework to aid exposure assessment of nanomaterials in consumer products. Ecotoxicology. 2008;17(5):438–47.

Higashisaka K, Yoshioka Y, Tsutsumi Y. Applications and safety of nanomaterials used in the food industry. Food Safety. 2015;3(2):39–47.

Kuzma J, Romanchek J, Kokotovich A. Upstream oversight assessment for agrifood nanotechnology: a case studies approach. Risk Anal. 2008;28(4):1081–98.

Mohammadlou M, Maghsoudi H, Jafarizadeh-Malmiri H. A review on green silver nanoparticles based on plants: synthesis, potential applications and eco-friendly approach. Int Food Res J. 2016;23(2):446–63.

Poland C. Nanoparticles: possible routes of intake. In: WHO Workshop on Nanotechnology and Human Health: Scientific Evidence and Risk Governance. Bonn, Germany; 2012. p. 10–11.

World Health Organization, Regional Office for Europe. Nanotechnology and human health: scientific evidence and risk governance. Copenhagen: WHO Regional Office for Europe. 2013.

Chapter 3
Novel Technologies in Food Nanobiotechnology

3.1 Introduction

Nanobiotechnology is the application of nanotechnology in biological fields, which is a multidisciplinary field that currently recruits approach, technology and facility available in conventional as well as advanced avenues of engineering, physics, chemistry and biology. Nanobiotechnology offers a wide range of opportunities for the development of innovative products and applications in the food system (Assa et al. 2016; de Morais et al. 2014). Self-cleaning surface, antibacterial coating and packaging (e.g. active and smart packaging), detecting food pathogens and monitoring of the food deterioration using nanosensors, delivery of nutraceuticals, functional ingredients, additives and farm food, food nanoemulsions and nanodispersions, enzyme immobilization and green synthesis of inorganic metals and metal oxides are the main parts of the food systems in which nanobiotechnology has a critical role in their structures and functionality (Vaghari et al. 2016). Enzyme immobilization, nanoemulsions/nanodispersions preparation and green synthesis of inorganic nanoparticles as three novel technology in food based on the nanobiotechnology are discussed in this chapter.

3.2 Enzyme Immobilization

Enzymes are ubiquitous natural biocatalysts which promote the transformation of chemical species in living systems. Enzymes, such as carbohydrates, proteases, lipases, lysozymes and oxidoreductases, are widely used in food industries due to their excellent catalytic activities (Assa et al. 2016). These biocatalysts are biocompatible, biodegradable and are derived from renewable resources. These molecules, consisting of thousands of atoms in precise arrangements, are able to catalyze the

H. Jafarizadeh-Malmiri et al., *Nanobiotechnology in Food: Concepts, Applications and Perspectives*, https://doi.org/10.1007/978-3-030-05846-3_3

multitude of different chemical reactions occurring in biological cells. Enzymatic processes are conducted under mild conditions (close to ambient temperature, atmospheric pressure and physiological pH) in water, with high rates and selectivity thus decreasing the formation of by-products. Practical use of enzymes has been realized in various industrial and biological processes. The application of enzymes due to their short lifetime, unstable structure and extensive separation costs are presently limited (Vaghari et al. 2016). Unlike more traditional organic and inorganic catalysts, enzymes are also large and fragile molecules. Instability and sensitivity to process conditions are disadvantages of using soluble enzymes. Therefore, application of solid phase biocatalysts has become more and more important during the last decades (Jafarizadeh-Malmiri et al. 2012).

There have been many approaches to improve the enzyme stability such as enzyme immobilization, enzyme modification, protein engineering and medium engineering. The immobilization of enzymes has proven particularly valuable and has been exploited over the last four decades to enhance enzyme properties such as activity, stability and substrate specificity for their successful utilization in industrial processes (Vaghari et al. 2016). Immobilized enzymes have a great importance in industrial bioprocesses, especially in food, nutritional and pharmaceutical technologies. There are several reasons for using an enzyme in an immobilized form. In addition to more convenient handling of the enzyme, it provides for its facile separation from the product. It helps to prevent the contamination of the substrate with enzyme/protein or other compounds, which decreases purification costs. Immobilization also facilitates the efficient recovery and reuse of costly enzymes, with longer half-life and less degradation, in successive batches, or the process can eventually be carried out in a continuously operating reactor (Sheldon 2007; Spahn and Minteer 2008).

Methods of enzyme immobilization can be divided into three main categories, namely binding to a support (carrier or matrix), entrapment (encapsulation) and cross-linking. The enzymes can be attached to the support via interactions ranging from reversible physical adsorption (such as hydrophobic and van der Waals interactions) and ionic linkages to stable covalent bonds. The strength of the binding is usually inversely related to the ease with which it can be reversed. The entrapment method is based on the occlusion of an enzyme within polymeric network typically organic or inorganic polymer matrices, such as polyacrylamide and silica sol–gel, respectively. The physical restraints generally are weak; therefore, the addition of covalent attachment is often required to prevent enzyme leakage. Cross-linking of enzymes employs a bifunctional cross-linking agent, most commonly glutaraldehyde in order to prepare carrier-free immobilized enzymes, such as cross-linked enzyme crystals and cross-linked enzyme aggregates. The molecular weight (MW) of the cross-linking agent is negligible compared with that of the enzyme, and the resulting biocatalyst essentially comprises a 100% active enzyme. So this approach offers clear advantages such as high enzyme activity in the catalyst and low production costs due to the exclusion of an additional (expensive) carrier (Sheldon 2007). Figure 3.1 shows enzyme immobilization methods.

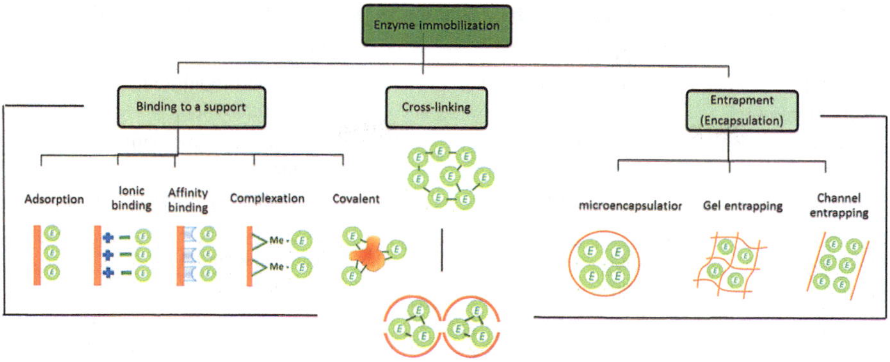

Fig. 3.1 Enzyme immobilization methods (Vaghari et al. 2016)

Enzymes can be immobilized via covalent attachment to stimulus-responsive or 'smart polymers' and magnetic particles. Consequently, particles can be separated from the reaction mixture by the means of magnetic decantation or magnetically stabilized fluidized bed reactors (Yiu and Keane 2012). Recent decades have witnessed the appearance of synthetic functional polymers which was originally called 'stimulus responsive' or 'smart' polymers (Galaev and Mattiasson 1999). Smart polymers, such as chitosan, are water-soluble polymers that undergo dramatic reversible changes in their solubility in response to an appropriate stimulus such as change of pH, ionic strength, electric or magnetic field, temperature, or addition of chemical species. In general, all smart polymers and hydrogel systems can be divided into five groups: pH-sensitive, heat-sensitive, phase sensitive, light sensitive smart polymers and polymers with dual stimuli-responsiveness (Ghaz-Jahanian et al. 2015).

The results of immobilization, including the performance of immobilized enzymes, strongly depend on the properties of supports, which are usually referred to as material types, compositions and structures. Under the scale of nano, nanomaterials have characteristics such as magnetism and large surface area to volume ratio. These characteristics are in favor of immobilization of the enzymes. Immobilization of enzymes onto biopolymers' nanoparticles, such as chitosan nanoparticles (NPs), has shown some benefits like improving their stability to pH and temperature, resistance to proteases and other denaturing compounds, as well as an adequate environment for their repeated use or controlled release (Assa et al. 2017). However, synthetic polymer, such as nanosilica aerogel, has been successfully used in enzyme immobilization, such as lipase (Amirkhani et al. 2016a, b). Table 3.1 indicates some of the immobilized enzymes onto the chitosan NPs.

Among various nanostructures, magnetic nanoparticles (MNPs) have a wide range of applications. They possess unique properties, namely superparamagnetism, low toxicity and biocompatibility. These properties are of use for a variety of potential applications such as immobilization of cells and enzymes, bioseparation systems, immunoassays, drug delivery and biosensors. The most commonly used

Table 3.1 Immobilized enzymes onto the chitosan NPs (Jafarizadeh-Malmiri et al. 2012)

Enzyme	Support	Preparation method
Alcohol dehydrogenase	Fe_3O_4–chitosan NPs	Covalent binding with glutaraldehyde
β-d-Galactosidase	Fe_3O_4–chitosan NPs	Covalent binding with glutaraldehyde
β-Galactosidase	Chitosan NPs	Precipitation, emulsion cross linking
Glucoamylase	Fe_3O_4–chitosan NPs	Ionic adsorption
Glucose oxidase	Chitosan NPs	Covalent binding
Laccase	Chitosan NPs	Reversed phase suspension
l-Laccase	Fe_3O_4–chitosan NPs	Ionic adsorption and covalent binding
Lipase	Fe_3O_4–chitosan NPs	Cross linking with trypolyphosphate
Neutral proteinase	Chitosan NPs	Ionic gelation
Pullulanase	Fe_3O_4–chitosan NPs	Photochemistry in aqueous suspension
Trypsin	Linolenic acid-modified chitosan NPs	Covalent binding with glutaraldehyde

magnetic particles are magnetite (Fe_3O_4) and maghemite (γ-Fe_2O_3). Magnetic nanosystems combining enzymes are one of the fastest growing areas of biotechnology research. A graphical overview of several types of MNP architectures with different core-shell arrangements is depicted in Fig. 3.2. The size of MNPs can vary from a few nm to several hundred nm in some condensed NP assemblies which exhibit diverse structural shapes (Vaghari et al. 2016). Some of immobilized enzymes onto MNPs are listed in Table 3.2.

3.3 Green Synthesis of Inorganic Nanoparticles

Green chemistry is defined as chemistry aided processes for pollution prevention which can be extended to specific areas including green analytical chemistry, environmentally friendly analytical chemistry and clean analytical methods. Incorporation of green chemistry techniques and methodologies into nanotechnology is of great interest which has gained much attention over the past decade. Green synthesis of metal and metal oxide nanoparticles (NPs), such as silver, gold, selenium, zinc oxide, titanium dioxide and iron oxide, have attracted considerable attention in recent years. In this regard, plant extracts and natural resources, such as microorganisms and enzymes, have been found to be good alternative reagents in NP synthesis. Utilizing green substances has several advantages including low energy consumption and moderate operation conditions (e.g. pressure and temperature) without using any toxic chemicals. As compared to the metallic elements in

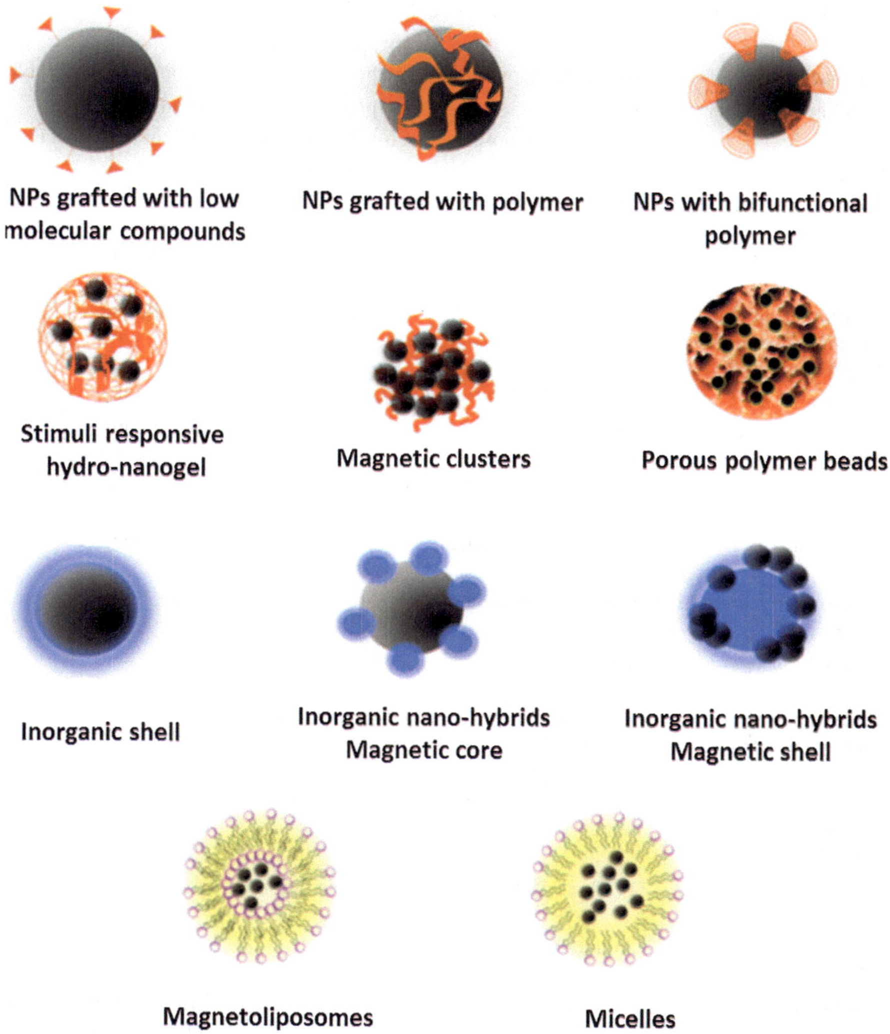

Fig. 3.2 General types of MNP arrangements with polymers, molecules and inorganic nanoparticles (Vaghari et al. 2016)

bulk state, metallic NPs exhibit unusual chemical, physical, biological, optical and thermal properties due to their high surface area to volume ratio. Therefore, these unique properties make NPs favorable for many different applications such as packaging, nanosensors and new antibiotics generation (Mohammadlou et al. 2016).

Generally, metal NPs can be produced using the top-down (physical) and bottom-up (chemical and biological) methodologies. Figure 3.3 indicates all top-down and bottom-up techniques used to synthesize metal NPs. The chemical synthesis methods are simple and easily controlled. However, they generate toxicity due to unwanted

Table 3.2 Some of the immobilized enzymes onto the MNPs (Vaghari et al. 2016)

Enzyme	Magnetic carrier	NPs preparation method	Immobilization method
Alkaline phosphatase	Fe_3O_4	Co-precipitation	Cross-linking
α-Amylase	Fe_3O_4 @ cellulose	Co-precipitation	Covalent attachment
Cellulase	Carboxylic acid functionalized Fe_3O_4	Co-precipitation	Cross-linking
Cholesterol oxidase	Gama-Fe_2O_3 @ SiO_2 @ APES	Co-precipitation	Cross-linking
D-Amino acid oxidase from *Rhodosporidium toruloides*	Fe_3O_4 @ APES	Co-precipitation	Cross-linking
Esterase	Fe_3O_4 @ APES	Co-precipitating	Cross-linking
Glucose oxidase	Amino-modified $CoFe_2O_4$ @ SiO_2	Co-precipitation	Cross-linking
β-glucosidase	Agarose-Fe_3O_4-ECH-IDA-Co^{2+}	Co-precipitation	Metal ion affinity
β-glucosidase	Sodium citrate coated Fe_3O_4	Co-precipitation	Cross-linking
Lipase from *Aspergillus niger*	Chitosan-coated Fe_3O_4	–	Covalent binding
Lipase from *Candida rugosa*	ɣ-Fe_2O_3		–
Lipase from *Semi tinmarcescens*	Aldehyde-functionalized Fe_3O_4 @ APES	Co-precipitation	–
Lipase from *Thermomyces lanuginosa*	Aminofunctionalized Fe_3O_4	Co-precipitation	Covalent binding
Papain	Fe_3O_4 @ SiO_2	Co-precipitation, Sol-gel	–
Trypsin	Amine-functionalized Fe_3O_4 nanoparticles	–	Cross-linking
Chitosanase	Fe_3O_4 @ amylose	–	Physical-adsorption and multipoint covalent bonding
Pectinase	AOT-Fe_3O_4	–	Co-precipitation

Note: APES: 3-aminopropyltriethoxysilane, AOT: Docusate sodium salt

harmful interactions with biological systems. Furthermore, in the chemical synthesis method, chemical reductants, such as sodium borohydride, and stabilizers, such as Polyvinylpyrrolidone, are widely used and their residual in the final products limit the applications of the fabricated NPs in medicine and food areas. Generally, metal NP synthesis has three steps, namely nucleation (by reducing the metal ions into the elements), growth of nucleuses and stabilizing of the formed NPs. Figure 3.4 shows the three steps of the silver NPs formation. Numerous physical methods (e.g. sonochemi-

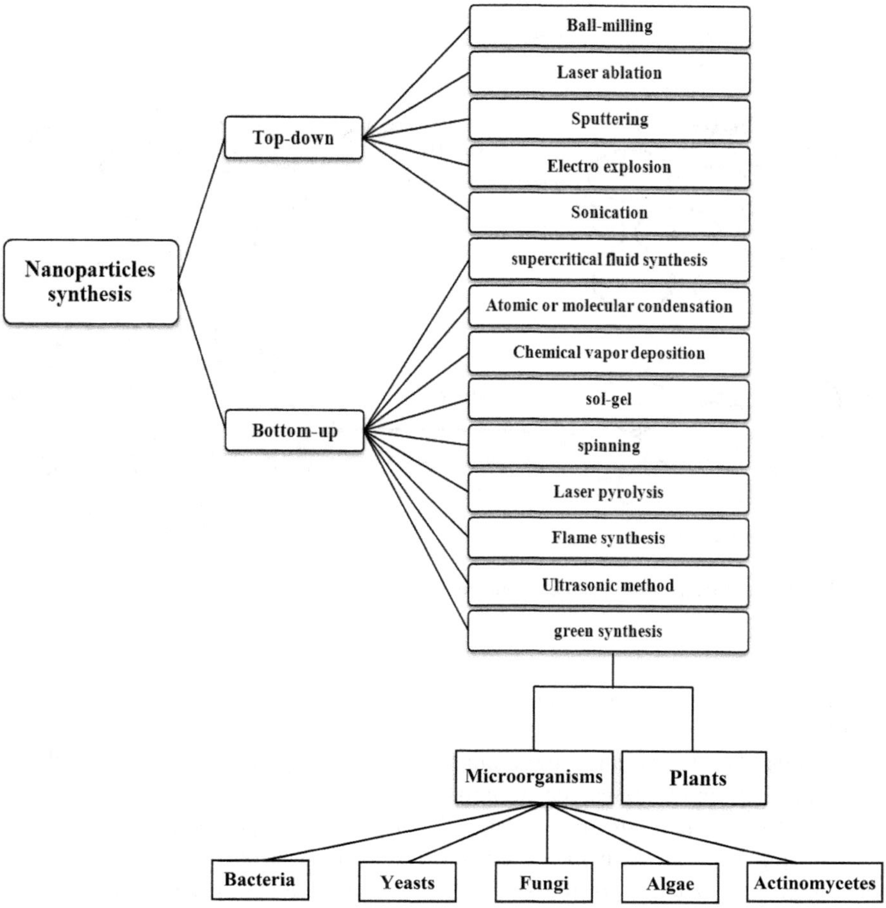

Fig. 3.3 Top-down and bottom-up techniques used to synthesize of metal NPs (Mohammadlou et al. 2016)

cal, ultrasound irradiation, ultraviolet (UV) irradiation, microwave and hydrothermal) have been applied to synthesize NPs. However, these techniques are expensive and unsustainable (Eskandari-Nojehdehi et al. 2016).

Metal NP synthesis using microorganisms such as bacteria, fungi, yeast and actinomycetes has immense potential and is an environmentally friendly process. Extracts from microorganisms, including enzymes, proteins, amino acids, polysaccharides and vitamins, may take part in NP synthesis as both reducing and capping agents. It seems that microorganisms have the potential to immobilize NPs by providing a viscous medium which, in turn, prevents their aggregation. Recently, these microorganisms have been known as possible eco-friendly nano-factories. Several studies have been reported successful on biological synthesis of silver NPs using microorganisms including *Verticillium* sp., *Aspergillus fumigatus*, *Aeromonas* sp., *Klebsiella pneumo-*

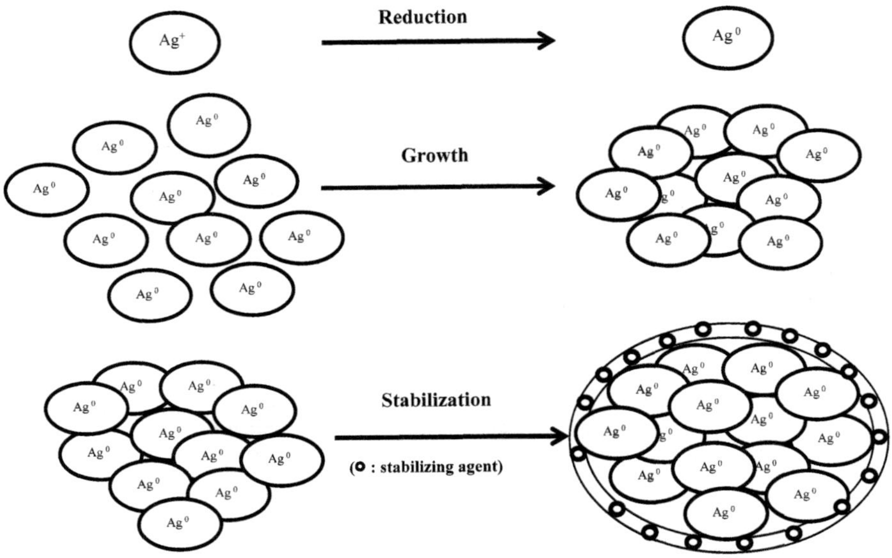

Fig. 3.4 Silver NPs formation steps (Mohammadlou et al. 2016)

nia, Escherichia coli, Enterobacter cloacae (Enterobacteriaceae), Aspergillus flavus and *Bacillus subtilis* (Mohammadlou et al. 2016).

Various plant metabolites, including terpenoids, polyphenols, sugars, alkaloids, phenolic acids and proteins play an important role in the bioreduction of metal ions to form NPs. Terpenoids are a group of diverse organic polymers synthesized in plants from five-carbon isoprene units which display strong antioxidant activity. Flavonoids are a large group of polyphenolic compounds that comprise several classes including anthocyanins, isoflavonoids, flavonols, chalcones, flavones, and flavanones, which can actively chelate and reduce metal ions into nanoparticles. Various functional groups of flavonoids are capable of forming NPs. Table 3.3 shows different bioreductant of numerous plants which have been used in the synthesis of metal NPs.

Metal NPs have a high surface energy, making them less stable. Therefore, resulting NPs aggregate and prefer to acquire a more stable morphology such as a truncated triangle to minimize their Gibbs free energy. Plant hydrocarbons, such as nonacosane and heptacosane, have a positive effect on stabilization of the formed NPs. In fact, the carbonyl group of amino acid, such as lysine, cysteine, arginine, and methionine residues and proteins has the potential to bind metal ions to form nanoparticles (e.g. capping of NPs) and to prevent agglomeration and thereby stabilize the medium (Ahmadi et al. 2018a; Mohammadlou et al. 2017). Table 3.4 indicates some of our previous studies on green synthesis of metal and metal oxide NPs using plant extracts and microorganisms.

Table 3.3 Plants and their bioreductants (Mohammadlou et al. 2016)

Plant	Bioreductant molecules
Jatropha curcas	Proteins
Carica papaya	Hydroxyflavones and catechins
Ocimum sanctum	Phenolic and flavonoid compounds, proteins, ascorbic acid, Gallic acid
Desmodium triflorum	Water-soluble antioxidative agents like ascorbic acids
Rosa rugosa	Carboxylate content, amine groups
Chenopodium album	Aldehyde, alkaloids, apocarotenoids, flavonoids
Acalypha indica	Flavonoids
Sesuvium portulacastrum	Proteins, flavones and terpenoids
Hibiscus rosa sinensis	Carboxylate ion groups
Achyranthes aspera	Polyols
Citrus sinensis	Vitamin C, flavonoids, acids and volatile oils
Mentha piperita	Menthol
Citrullus colocynthis	Polyphenols with aromatic ring and bound amide region
Anacardium occidentale	Polyols and proteins
Zingiber officinale	Alkanoids, flavonoids
Piper betle	Proteins
Solanum xanthocarpum	Phenolics, alkaloids and sugars
Glycyrrhiza glabra	Flavonoids, terpenoids, thiamine
Piper nigrum	Proteins
Trianthema decandra	Hydroxyflavones and catechins
Dioscorea bulbifera	Polyphenols or flavonoids
Elettaria cardamomum	Alcohols, carboxylic, acids, ethers, esters and aliphatic amines
Leonuri herba	Polyphenols and hydroxyl groups
Morinda pubescens	Hydroxyflavones, catechins
Olibanum	Hydroxyl, carbonyl
Annona squamosa	Sugars (aldoses) and terpenoids
Piper betle	Proteins
Plumeria rubra	Proteins
Hydrilla verticillata	Proteins
Lantana camara	Carbohydrates, glycosides and flavonoids
Andrographis paniculata	Hydroxyflavones catechins
Annona squamosa	Glycoside, alkaloids, saponins, flavonoids, tannins, carbohydrates, proteins, phenolic compounds, phytosterols, and amino acids
Malva parviflora	Proteins
Hibiscus cannabinus	Ascorbic acid
Castor oil, Khat and Sun flower	Proteins, phenols and flavonoids, terpenoids
Artocarpus heterophyllus	Amino acids, amides group

(continued)

Table 3.3 (continued)

Plant	Bioreductant molecules
Cocos nucifera	Hydrocarbon such as nonacosane and heptacosane
Tithonia diversifolia	Proteins, polysaccharides, terpenoids
Coleus aromaticus	Flavonoids
Mango	Aldehydes, ketone, carboxyl and hydroxyl
Aloe	Protein, alkaloids, flavonoids,
Syzygium cumini	Flavonoids
Tea	Amides, carboxyl, amino groups and poly phenols
Jatropha curcas	Proteins
Carica papaya	Hydroxyflavones and catechins
Ocimum sanctum	Phenolic and flavonoid compounds, proteins ascorbic acid, Gallic acid, terpenoids,
Desmodium triflorum	Water-soluble antioxidative agents like ascorbic acids

Table 3.4 Some of our newest studies on green synthesis of metal and metal oxide NPs

Metal/metal oxide NPs	Reducing and stabilizing agents	Reference
Gold	*Agaricus bisporus* extract	Eskandari-Nojehdehi et al. (2016)
Silver	*Pelargonium* leaf extract	Mohammadlou et al. (2017)
Silver	Chitosan	Torabfam and Jafarizadeh-Malmiri (2018)
Silver	*Aspergillus fumigatus*	Ghanbari et al. (2018)
Gold	*Agaricus bisporus* extract	Eskandari-Nojedehi et al. (2018)
Gold	Arabic gum	Eskandari-Nojehdehi et al. (2018)
Silver	*Aloe vera* leaf extract	Ahmadi et al. (2018a)
Silver	*Aloe vera* leaf extract	Ahmadi et al. (2018b)
Silver	*Juglans regia* leaf extract	Eshghi et al. (2018)
Selenium	*Pelargonium* leaf extract	Fardsadegh et al. (2018)
Zinc oxide	*Pelargonium* leaf extract	Vahidi et al. (2018)

3.4 Nanoemulsions

Colloidal emulsions have gained much more attention in the food, pharmaceutical, agrochemical and other industries due to their ability to produce lipid-soluble nutraceuticals such as antimicrobial and antioxidant agents, drugs and vitamins. In fact, the emulsions, with demonstrated health benefits and using colloidal dispersions to encapsulate, protect and deliver bioactive lipophilic components. Recent progress in nanotechnology has also attracted considerable attention for improving the water solubility and bioavailability of lipophilic bioactive compounds. Preparation of functional lipids in the forms of nanoemulsions is expected to greatly improve their solubility and bioavailability (Sayyar et al. 2017).

Nanoemulsions are transparent or translucent, colloidal dispersions, usually in the 50–1000 nm size range. There are two basic types of nanoemulsions most likely to be formed depending on the composition that is oil-in-water (O/W), wherein oil droplets are dispersed in the continuous aqueous phase or water-in-oil (W/O).

Multiple emulsions are termed 'emulsions of emulsions', the droplets of the dispersed phase themselves containing even smaller dispersed droplets. Each dispersed globule in a water-in-oil-in-water W/O/W double nanoemulsion forms a vesicular structure with single or multiple aqueous compartments separated from the aqueous dispersion phase by a layer of oil phase compartments. Double emulsions may offer some advantages for food application mainly with relation to their capability to encapsulate (Anarjan et al. 2014; Garti 1997; McClements 2012; Sayyar et al. 2017). Figure 3.5 indicates the schematic of various nanoemulsions.

Reducing particle size in emulsion-based delivery systems has a number of consequences that may be beneficial for certain food and pharmaceutical applications. Table 3.5 shows a comparison between nanoemulsions and microemulsions.

Nanoemulsions containing lipophilic bioactive substances can be produced using various emulsification processes. Nowadays, several methods for preparing nanoemulsions are available. They can be produced using both low-energy, in which a change of curvature and a phase transition takes place during the emulsification process, and high-energy protocols that consist of the application of high mechanical energy during emulsification (Anarjan et al. 2015). These methods are included: high pressure homogenization, microfluidization, sonication, phase inversion temperature, phase inversion composition, spontaneous emulsification and solvent displacement (Sayyar et al. 2017). Table 3.6 provides a summary of the prepared nanoemulsions using different methods and emulsifiers.

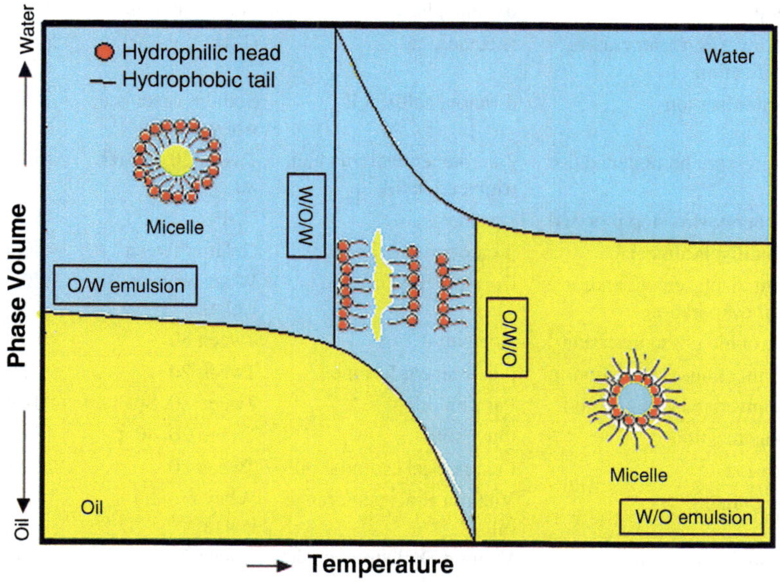

Fig. 3.5 Schematic phase diagram of different nanoemulsions (Sayyar et al. 2017)

Table 3.5 Comparison between nanoemulsions and microemulsions (Sayyar et al. 2017)

Properties	Nanoemulsions	Microemulsions
Appearance	Transparent/translucent with bluish coloration/higher optical clarity	Transparent
Droplets diameter	50–1000 nm	10–120 nm
Structure system	Spherical droplets	Hexagonally packed columns/ lamellar sheets/worm like micellar phases/spherical droplets
Surface area	Higher surface area and free energy	Lower surface area
Concentration of surfactant	Low	High
Method of preparation	Low and high energy methods	Spontaneous emulsification
Stability	Kinetically stable/metastable, thermodynamically unstable	Thermodynamically stable
Bioavailability	High oral bioavailability	Low oral bioavailability
Agglomeration of droplets	Greater stability to droplet aggregation and gravitational separation/They do not show the problems of inherent creaming, flocculation, coalescence and sedimentation	Droplet aggregation and gravitational separation/They show the problems of coalescence and sedimentation

Table 3.6 Some prepared nanoemulsions using different methods (Anarjan et al. 2015)

Method	Bioactive component/ organic phase	Surfactant type	Droplet size (nm)
Phase inversion temperature emulsification	Hexadecane/–	$C_{12}E_4$	50–130
Microfluidization	d-limonene/fish oil	Sodium caseinate, whey protein	<150
High pressure homogenization	β-carotene/medium chain triglyceride oil	Tween 20, 40, 60, 80	40–400
Low-energy two-step process	Decane/–	$C_{12}E_5$	15
Solvent displacement	β-carotene/acetone	Gelatin/Tween	<80
Solvent displacement and/or solvent evaporation	Corn oil/ethyl acetate	Whey protein isolate	<70
Catastrophic phase inversion	Acetem/–	Tween 60	<100
Emulsification–evaporation	β-carotene/n-hexane	Tween 20	9–280
Phase inversion composition	Paraffin oil/–	Tween 80, Span 80	50–300
Homogenization	Olive oil/–	Tween 20, 40, 6	45–500
Sonication	Orange peel essential oil/–	Tween 80	12.68
Homogenizing	Vitamin E-acetate/orange oil	Whey protein isolate	110
Spontaneous emulsification	Vitamin D/medium chain triglycerides	Tween 20, 40, 60, 80, 85	<200

As clearly observed in Table 3.6, the main components of the nanoemulsion process are a mixture of oil, water and surfactant. The surfactant is used to create a low interfacial tension, which aids in the production of nanosized particles. Biosurfactants are surface active substances that reduce interfacial tension and are produced or excreted at the microbial cell surface. Rhamnolipid, as a biosurfactant, has been derived from *Pseudomonas aeruginosa*. Surfactin was originally obtained from the cell-free broth of *Bacillus subtilis*. Xanthan gum has been achieved by *Xanthomonas campestris* (de Morais et al. 2014). All these biosurfactants have potential in preparation of the food nanoemulsions.

References

Ahmadi O, Jafarizadeh-Malmiri H, Jodeiri N. Eco-friendly microwave-enhanced green synthesis of silver nanoparticles using Aloe vera leaf extract and their physico-chemical and antibacterial studies. Green Process Synth. 2018a;7(3):231–40.

Ahmadi O, Jafarizadeh-Malmiri H, Jodeiri N. Optimization of processing parameters for hydrothermal silver nanoparticles synthesis using Aloe vera leaf extract and estimation of their physico-chemical and antifungal properties. Z Phys Chem. 2018b. https://doi.org/10.1515/zpch-2017-1089.

Amirkhani L, Moghaddas J, Jafarizadeh-Malmiri H. Candida rugosa lipase immobilization on magnetic silica aerogel nanodispersion. RSC Adv. 2016a;6(15):676–12687.

Amirkhani L, Moghaddas J, Jafarizadeh H. Optimization of Candida rugosa lipase immobilization parameters on magnetic silica aerogel using adsorption method. Iranian J Chem Eng. 2016b;13:19–31.

Anarjan N, Nehdi IA, Sbihi HM, Al-Resayes SI, Malmiri HJ, Tan CP. Preparation of astaxanthin nanodispersions using gelatin-based stabilizer systems. Molecules. 2014;19(9):14257–65.

Anarjan N, Jafarizadeh-Malmiri H, Nehdi IA, Sbihi HM, Al-Resayes SI, Tan CP. Effects of homogenization process parameters on physicochemical properties of astaxanthin nanodispersions prepared using a solvent-diffusion technique. Int J Nanomedicine. 2015;10:1109.

Assa F, Jafarizadeh-Malmiri H, Ajamein H, Anarjan N, Vaghari H, Sayyar Z, Berenjian A.A biotechnological perspective on the application of iron oxide nanoparticles. Nano Res. 2016;9(8):2203–25.

Assa F, Jafarizadeh-Malmiri H, Ajamein H, Vaghari H, Anarjan N, Ahmadi O, Berenjian A. Chitosan magnetic nanoparticles for drug delivery systems. Crit Rev Biotechnol. 2017;37(4):492–509.

de Morais MG, Martins VG, Steffens D, Pranke P, da Costa JAV. Biological applications of nanobiotechnology. J Nanosci Nanotechnol. 2014;14(1):1007–17.

Eshghi M, Vaghari H, Najian Y, Najian M, Jafarizadeh-Malmiri H, Berenjian A. Microwave-assisted green synthesis of silver nanoparticles using Juglans regia leaf extract and evaluation of their physico-chemical and antibacterial properties. Antibiotics. 2018;7(3):68.

Eskandari-Nojedehi M, Jafarizadeh-Malmiri H, Rahbar-Shahrouzi J. Hydrothermal green synthesis of gold nanoparticles using mushroom (Agaricus bisporus) extract: physico-chemical characteristics and antifungal activity studies. Green Process Synth. 2018;7(1):38–47.

Eskandari-Nojehdehi M, Jafarizadeh-Malmiri H, Rahbar-Shahrouzi J. Optimization of processing parameters in green synthesis of gold nanoparticles using microwave and edible mushroom (Agaricus bisporus) extract and evaluation of their antibacterial activity. Nanotechnol Rev. 2016;5(6):537–48.

Eskandari-Nojehdehi M, Jafarizadeh-Malmiri H, Jafarizad A. Microwave accelerated green synthesis of gold nanoparticles using gum Arabic and their physico-chemical properties assessments. Z Phys Chem. 2018;232(3):325–43.

Fardsadegh B, Vaghari H, Mohammad-Jafari R, Najian Y, Jafarizadeh-Malmiri H. Biosynthesis, characterization and antimicrobial activities assessment of fabricated selenium nanoparticles using Pelargonium zonale leaf extract. Green Process Synth. 2018. https://doi.org/10.1515/gps-2018-0060.

Galaev IY, Mattiasson B. 'Smart' polymers and what they could do in biotechnology and medicine. Trends Biotechnol. 1999;17(8):335–40.

Garti N. Progress in stabilization and transport phenomena of double emulsions in food applications. LWT-Food Sci Technol. 1997;30(3):222–35.

Ghanbari S, Vaghari H, Sayyar Z, Adibpour M, Jafarizadeh-Malmiri H. Autoclave-assisted green synthesis of silver nanoparticles using A. fumigatus mycelia extract and the evaluation of their physico-chemical properties and antibacterial activity. Green Process Synth. 2018;7(3):217–24.

Ghaz-Jahanian MA, Abbaspour-Aghdam F, Anarjan N, Berenjian A, Jafarizadeh-Malmiri H.Application of chitosan-based nanocarriers in tumor-targeted drug delivery. Mol Biotechnol. 2015;57(3):201–18.

Jafarizadeh-Malmiri H, Ghaz-Jahanian MA, Berenjian A. Potential applications of chitosan nanoparticles as novel support in enzyme immobilization. Am J Biochem Biotechnol. 2012;8(4):203–19.

McClements DJ. Crystals and crystallization in oil-in-water emulsions: implications for emulsion-based delivery systems. Adv Colloid Interface Sci. 2012;174:1–30.

Mohammadlou M, Maghsoudi H, Jafarizadeh-Malmiri H. A review on green silver nanoparticles based on plants: synthesis, potential applications and eco-friendly approach. Int Food Res J. 2016;23(2):446–63.

Mohammadlou M, Jafarizadeh-Malmiri H, Maghsoudi H. Hydrothermal green synthesis of silver nanoparticles using Pelargonium/Geranium leaf extract and evaluation of their antifungal activity. Green Process Synth. 2017;6(1):31–42.

Sayyar Z, Anarjan N, Jafarizadeh Malmiri H, Berenjian A. Nanoemulsions: preparation, characterization, applications, and their kinetic and thermodynamic stability. In: Acosta MJ, editor. Advances in energy research. New York: Nova Science; 2017. p. 1–56.

Sheldon RA. Enzyme immobilization: the quest for optimum performance. Adv Synth Catal. 2007;349(8–9):1289–307.

Spahn C, Minteer SD. Enzyme immobilization in biotechnology. Recent Patents Eng. 2008;2(3):195–200.

Torabfam M, Jafarizadeh-Malmiri H. Microwave-enhanced silver nanoparticle synthesis using chitosan biopolymer: optimization of the process conditions and evaluation of their characteristics. Green Process Synth. 2018;7:530–37.

Vaghari H, Jafarizadeh-Malmiri H, Mohammadlou M, Berenjian A, Anarjan N, Jafari N, Nasiri S.Application of magnetic nanoparticles in smart enzyme immobilization. Biotechnol Lett. 2016;38(2):223–33.

Vahidi A, Vaghari H, Najian Y, Najian MJ, Jafarizadeh-Malmiri H. Evaluation of three different green fabrication methods for the synthesis of crystalline ZnO nanoparticles using Pelargonium zonale leaf extract. Green Process Synth. 2018. https://doi.org/10.1515/gps-2018-0097.

Yiu HH, Keane MA. Enzyme–magnetic nanoparticle hybrids: new effective catalysts for the production of high value chemicals. J Chem Technol Biotechnol. 2012;87(5):583–94.

Chapter 4
Nano-additives for Food Industries

4.1 Introduction

Functional ingredients such as vitamins, antimicrobials, antioxidants, flavorings, colorants and preservatives in different molecular and physical forms, such as polarities (e.g. polar, nonpolar, amphiphilic), have been widely utilized in food systems and packaging. The nanostructured nutraceuticals and bioactive compounds incorporated into nano-sized delivery systems are usually known as nanoadditives. These nanomaterials are designed to be added in various food and beverage systems or their packaging polymers matrix, to promote their characteristics and give them various activities, which are described briefly in this chapter.

The unique characteristics of nano-materials provide many novel prospects in the food industries. The food systems can hold various natural nano-sized compounds like fats and carbohydrates as well as synthetic nanomaterials such as nano-emulsion, nanoencapsulated functional lipid compounds or their nanoparticles, nano-fibers, nano-liposomes etc. Moreover, the used nano-materials in food and beverage formulations or their packaging systems have either inorganic or organic structures (Bajpai et al. 2018).

The main reason for incorporation of nanomaterials in food formulations can be related to their great solubility and consequently the bioavailability enhancement as well as the protection ability of bioactive compounds during either process or storage of foods (Tan and Nakajima 2005). Various research reports that the nanomaterials can effectively enhance the quality of food by increasing the nutritional value, improving the organoleptic characteristics, boosting safety due to their antimicrobial and antioxidant activities, raising the packaging efficacy, and shelf life extension etc. (Nickols-Richardson 2017). However, using an economical operations technique as well as non-toxic, biocompatible and safe nano-delivery systems are some challenges in applications of nano-additives in food industries (Bajpai et al. 2018).

H. Jafarizadeh-Malmiri et al., *Nanobiotechnology in Food: Concepts, Applications and Perspectives*, https://doi.org/10.1007/978-3-030-05846-3_4

4.2 Nanomaterial Classifications

Nanomaterials would be mostly used in food formulations and food packaging. However, researchers have recently reported the beneficial uses of nanomaterials in food processing and food safety issues, as well. Furthermore, nano-sensors can be incorporated into the packaging polymers to provide various smart packaging and enhance the foods' safety (Chaudhry et al. 2008). Using the nano-filtrations in food processes, such as elimination of detrimental constituents from products, or the nano-fluids for thermal processes of the foods such as sterilization, are some applications of nanomaterials in food processes.

The foods with nanomaterial ingredients mainly offer improved, homogeneous and stable texture, pleasant taste, desired appearance and healthier than conventionally formulated products without nanomaterial constituents. Nanoemulsions, microemulsions, liposomes, nanoparticles, transferosomes, ethosomes and biopolymer-based delivery organizations are some common nutraceutical and supplement nano-encapsulating systems in food formulation uses. Nanoencapsulation is the second largest area (after food packaging) of application of nanotechnology in the food industry (He and Hwang 2016; Martirosyan and Schneider 2014). The nano-sized nutraceutical elements of food formulation can change the circulation of the constituents through the body. Thus, if a nano-sized delivery system can carry the bioactive element into the blood flow, its absorption, distribution, digestion and secretion would be different from its conventional forms. However, the broken nano-sized delivery systems, especially at the gastrointestinal region, cannot alter the absorption and distribution of delivered nutraceuticals from their conventional forms (Bajpai et al. 2018; Chaudhry et al. 2008).

According to Nano-forum, 'Nano-food' is the food that the nanotechnology procedures or nano-sized compounds are used during their cultivation, manufacture, formulation or packaging (Sekhon 2010; Sozer and Kokini 2009). The nano-food products would be classified into various classes, which are described below.

4.2.1 Inorganic Nanomaterials

Inorganic materials have found various uses in either food formulation, food packaging or food processing. For example, calcium, magnesium, selenium and iron nanoparticles can be used in food or beverage systems as a nutrition enhancer or preserver. Gold, silver, iron, silica as well as metal oxide nanoparticles, such as titanium oxide, zinc oxide and iron oxide, with antimicrobial, antioxidant and anti-odorant abilities, can also be used in food packaging constructions, giving effective active packaging. Silver, magnesium oxide and zinc oxide nanoparticles are the most efficient nanoparticles with antimicrobial activity. Titanium dioxide nanoparticles can protect the transparent polymers from UV radiations. Titanium nitride nanoparticles can enhance the mechanical strength of the polymeric packaging.

Thus, the bio-polymers can be used as an alternative for fossil-fuel based polymers in food packaging vector (Palza 2015).

Nanotechnology affords the manufacture of smart packaging, as well as containing metal based nanosensors to give valuable information about their inside packaged foods, in which it can sense, detect and visualize the food spoilage compounds (Bajpai et al. 2018; Chaudhry et al. 2008). For instance, TiO_2 nanoparticles with light sensibility can be incorporated into oxygen sensing ink, and as a result it can detect oxygen in exposure to UV light. The graphene oxide-TiO_2 nanoparticles were also developed as a photocatalyst for UV-activated oxygen detection (Son et al. 2015). Conductive inks based on copper nanoparticles with sizes of 40–50 nm for ink jet printing have also been developed by Park et al. (2007). This attractive and low-cost process is a substitute to expensive conductive metals such as gold. Nanosensors, inserted into food products as small chips, are imperceptible to the human eye and can also act as microelectronic barcodes (Bajpai et al. 2018; Chaudhry et al. 2008; He and Hwang 2016). Some inorganic nanoparticles can immobilize the enzymes which are used during food processing in order to improve their dispersion all through the food matrices, increase their activity and re-usability. For example, the activity, stability and adaptability of porcine triacylglycerol lipase, which are used in the hydrolyzing of olive oil, effectively increased as it immobilized covalently on silicon dioxide nanoparticles (Chaudhry et al. 2008; He and Hwang 2016).

4.2.2 Surface Functionalized Nanomaterials

Surface functionalized nanomaterials would be achieved by surface modification of nanoparticles using either hydrophilic or mucoadhesive emulsifiers or biopolymers, leading to a considerable amendment of the zeta potential, hydrophilicity, stability and adhesive and cellular uptake of nanoparticles (Gunasekaran et al. 2014). Surface functionalized nanomaterials are quit advance engineering nanomaterial, which can be used either in food formulation or food packaging uses. They give specific function, such as preservation or antimicrobial activity, to food matrix through oxygen absorption (Anarjan and Tan 2013a; Huang et al. 2010). In food packaging, the surface functionalized nanomaterials can bind to packaging polymers in order to improve their gas barrierity, moisture or volatile components diffusivity and mechanical strength. Despite inert nanomaterials they are able to bond to the polymer and are mostly not free enough to migrate from packaging into food matrix, or location into other tissues than the gastrointestinal region (Tsagkaris et al. 2018; Youssef and El-Sayed 2018). The nano-clay, which is mainly the montmorillonite with nano-scaled layers, is the most suggested nanosystem, which can bind to the packaging polymer matrix, making the organic/inorganic hybrids, and modify the polymer characteristics such as gas barrierity (Youssef and El-Sayed 2018). The antimicrobial activity can also be introduced to the biodegradable packaging using the surface functionalized nanosystems. For example, recently scientists have been bonded the benzoic acid to Mg–Al hydrotalcite and combined them into polycaprolactone, in order to provide an active antimicrobial food packaging (He and Hwang 2016; Sekhon 2010).

4.2.3 *Organic Nanomaterials*

The organic nanosystems, such as nano-sized vitamins and antioxidants, can also be designed to be used in food sectors in order to increase their cellular uptake and bio-absorption and enhance their bioavailabilities, compared to conventional bulk equivalents. It was predicted that if the food additives, such as preservations (e.g. citric acid, benzoic acid phosphoric acid) or nutraceuticals (e.g. vitamins A, D and E, carotenoids, poly-unsaturated fatty acids, and Co-Q_{10}), colorants etc. would be available as nano-sized, and their functionality can be considerably increased (Sekhon 2010). For example, in our previous research, the astaxanthin (a carotenoid with high antioxidant activity) nanodispersions have been formulated using various stabilizers. It was concluded that the gained nano-sized astaxanthin showed very increased cellular uptake as well as antioxidant activities than their micro-sized equivalents (Anarjan et al. 2012). The similar results have been reported for other nano-sized functional lipid compounds' nanoparticles. Recently, the nanoparticles, such as fullerenes and carbon nanotubes, have also been used in food processes as an absorbent to remove the undesired odor, colour, pesticide residuals as well as to clarify them. These nano-absorbents showed considerable efficiency in absorbing the organic materials due to their great surface area, resulting from their size reductions into nano-ranges (Hou et al. 2014; Paszkiewicz et al. 2016).

Furthermore, the organic nanoparticles, such as carbon nanotubes, can also be incorporated into the food packaging matrix in order to improve their mechanical strength and elasticity, flexibility, gas barrierity or adjust their electrical conductivities. The essential oil nanoparticles with antioxidant and antimicrobial activities can also be incorporated into packaging materials to give various active packaging. If the nanosensors (nanobiosensors), such as immobilized enzymes, are incorporated to the packaging polymer, it offers smart packaging (Bajpai et al. 2018; He and Hwang 2016; Martirosyan and Schneider 2014).

4.2.4 *Nanoencapsulated Compounds*

Nanoencapsulation is the technique to pack the constituents in order to protect them against environmental destructing stresses, cover their undesired characteristics, such as odor or reactivity, and control their release. It can be organized in nanocomposite, nanoemulsions, liposomes and nano-blended forms (de Souza Simões et al. 2017; Katouzian and Jafari 2016). Thus the protection, activity enhancement and stabilization of bioactive compounds, such as proteins, lipids, polysaccharides, vitamins and antioxidants, can be attained using nanoencapsulation processes (Prakash et al. 2018). The different processes are suggested to provide these compounds' nanocapsules in various studies, including high or low energy input techniques (Anarjan et al. 2011b). Thus, based on novel nanoencapsulation and control release techniques, different delivery systems can be designed to increase the cell absorption of valuable health promoting compounds and, consequently, decreases the amount of needed ingredient. Therefore, nanoencapsulation can give considerable savings in which the nano-sized compounds had significant increased surface-based characteristics, such as solubility, conductivity and colorant activity etc., in comparison to their micro-sized form (de Souza Simões et al. 2017; Prakash et al. 2018).

Several synthetic or natural surface active compounds, in either small molecular or polymer forms, have the potential to be used as emulsifiers, covering molecules and stabilizers in the nanoencapsulation of nutraceuticals. Thus, the proteins and polysaccharides are desired to be used in nanoencapsulation formulations (Anarjan et al. 2017; Anarjan and Tan 2013b). The protein–carbohydrate conjugations with improved stability, water-solubility and emulsifying properties, increased antioxidant activity and reduced allergenicity. They have been used in oil compounds' nanoencapsulation like essential oils fish oils and polyunsaturated fatty acids (Augustin et al. 2006; Shah et al. 2012).

It was confirmed that the nanoencapsulation of essential oils could enhance their antimicrobial activities and, consequently, their preservation efficiency in the food system. For example, nanoencapsulated terpenes and D-limonene, Mentha piperita, Cardamom, thymol and carvacrol essential oil showed better antibacterial activity against various foodborne pathogens than their micro-sized equivalents (Prakash et al. 2018).

The volatile compounds, which provide the food's flavor, taste, odor and aroma greatly affect their desirability by consumers. However, they have a tendency to decline because of evaporation, oxidation and degradation. Thus, their retention techniques are interesting for food scientists. Nanoencapsulation of these compounds can protect them and control their releases, effectively meeting the food industry challenges about the extension their shelf life, freshness, long lasting flavor and organoleptic features (Katouzian and Jafari 2016). Contrary, some healthy components, such as fish oil, have an undesirable taste or odor. Nanoencapsulation techniques can also resolve this problem by masking the unpleasant tastes and odors. The nanoencapsulated essential oil can be released by diffusion, wall material degradation or bursting the nanocapsules by swelling (Osorio-Tobón et al. 2016).

4.3 Nanostructures' Food Additives

The nanostructured nutraceuticals and bioactive compounds incorporated into nano-sized delivery systems are usually known as nanoparticles. These nanoparticles are designed to be added in various food and beverage systems or their packaging polymers matrix to promote their characteristics and give them various activities, which are described briefly below.

4.3.1 Solubility

The water insolubility of the most important nutraceuticals, such as fat-soluble vitamins, carotenoids, phytosterols, essential oils and poly unsaturated fatty acids etc., make their uses problematic, especially in aqueous-based food formulations. Moreover, various researchers have revealed the close relationship between the solubility and cellular uptake of nutritional compounds. Size reduction of the compounds, especially into nano-ranges, increases their saturation solubility as well as solubilizing rate, due to their considerable upturned surface area. Thus, nanotechnology offers homogeneous dispersion of these functional lipid bioactives into water, using nano-sized delivery systems.

According to literature, various studies have been conducted in order to incorporate these lipophilic bioactives into nano-sized delivery systems which are dispersible in water systems. Microemulsion liposomes, nanoemulsions/nanodispersions and solid lipid nanoparticles are the most promising delivery systems with water-solubility increasing uses.

Table 4.1 listed some conducted studies in preparation of water dispersed functional lipid nanostructures for food formulation uses. As can be seen in Table 4.1, various research groups all over the world are trying to formulate several water dispersible nanoparticles of nutraceuticals in order to increase their bioavailability and cellular uptake and comfort their uses into aqueous food systems. The incorporation of lipophilic nutrients into emulsifiers or a surface active biopolymer-based delivery system which can be used in numerous food systems are shown in Fig. 4.1.

4.3.2 Bioavailability

The bioavailability is the portion of an ingested constituent that finally ends up in the systemic flow (Holst and Williamson 2008). Most of the nutraceuticals, especially the functional lipid compounds, have less water bioavailability, despite notable *in-vitro* activities, thanks to their poor solubility or inappropriate molecular sizes. Enzymatic and chemical barriers in the gastrointestinal zone hinder the bioavailabilities of bioactive compounds. The gastrointestinal epithelium causes less permeability of numerous nutraceuticals, as well. The insolubility of functional lipid substitutes in the gastrointestinal tract is another reason for less bioavailability of these compounds. Thus, the increasing of the bioavailability of nutraceuticals using various nano-sized delivery systems, for example Phospholipid-based systems, have been proposed by various researchers (Kesarwani and Gupta 2013). Several studies indicated that the bioavailability of functional lipid bioactive

Table 4.1 Some water dispersible nano-sized functional lipid compounds

Functional lipid compounds	Delivery system	Emulsifiers or stabilizers	References
β-Carotene	Nanodispersions Nanoemulsions Microemulsions Solid lipid nanoparticles	Polyglycerol esters of fatty acids Modified starch, etc.	Anarjan et al. (2017), de Paz et al. (2012), Qian et al. (2013), Ribeiro et al. (2008), Silva et al. (2011), Tan and Nakajima (2005), Wang et al. (2012), Yin et al. (2009)
Lycopene	Nanodispersions Solid lipid nanoparticles	Tween 20, Tween 80, Lecithin, Sodium caseinate and gum Arabic	Anarjan and Jouyban (2017), dos Santos et al. (2015), Shariffa et al. (2016, 2017)
Lutein	Nanodispersion Nanoemulsions Solid lipid nanoparticles	Tween 80, Sodium dodecyl sulfate, Sodium caseinate and their mixture	Surh et al. (2017), Tan et al. (2016a, b), Teo et al. (2017)
Astaxanthin	Nanodispersions	Tween 20, Tween 40, Tween 60, Tween 80 Sodium caseinate, gum Arabic, pectin, Gelatin, mixed emulsifiers and biopolymers Other polysaccharides and etc.	Anarjan et al. (2010, 2011a, b, 2012, 2014), Anarjan and Tan (2013a, b)
Vitamin E (or tocopherols)	Nanodispersions Nanoemulsions Microemulsions	Tween 20, Quillaja saponins, Whey protein isolate, gum Arabic, etc.	Cheong et al. (2008), Mayer et al. (2013), Ozturk et al. (2015a), Saberi et al. (2014), Xu et al. (2018)
Vitamin D	Nanoemulsions Microemulsions	Tween 20, Soybean lecithin and their mixtures, Quillaja saponin, etc.	Golfomitsou et al. (2018), Kadappan et al. (2018), Ozturk et al. (2015b)
Coenzyme-Q10	Nanoemulsions	Tween 20	Sun et al. (2012)
Polyunsaturated fatty acids (ω3, DHA, etc.)	Nanoemulsions Microemulsions	Tween 20, Tween 80, Tween 40, Sodium caseinate, Soya lecithin, Sunflower phospholipids, etc.	Chen et al. (2017), Esquerdo et al. (2018), Karthik and Anandharamakrishnan (2016), Walker et al. (2017)
Various essential oils	Nanoemulsions Solid lipid nanoparticles	Tween 20, Tween 40, Tween 80 Various polysaccharides such as pectin and etc.	Jemaa et al. (2018), Prakash et al. (2018), Sotelo-Boyás et al. (2017)

compounds, such as Co-Q10, α-tocopherol, phytosterol, polyunsaturated fatty acids, β-carotene, astaxanthin and curcumin etc., have been considerably increased by their incorporation into nano-sized delivery systems (Anarjan et al. 2011b, 2012; Augustin and Sanguansri 2009; Huang et al. 2010; Kesarwani and Gupta 2013).

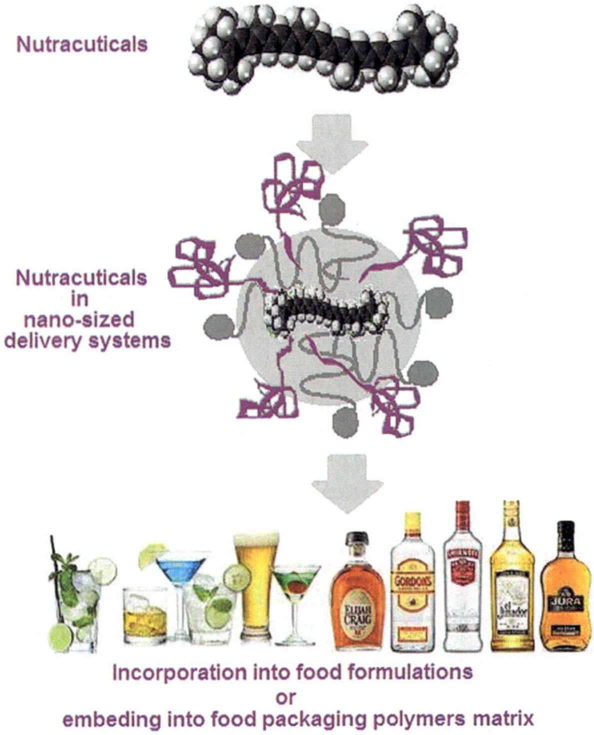

Fig. 4.1 Schematic delivery of functional lipid (hydrophobic) bioactive components by hydrophilic nano-sized delivery systems (which are formulated by surface active small molecular compounds or biopolymers) in food matrix

Similar results have also been observed for iron, calcium and selenium etc., in which the metal nanoparticles have enhanced bioavailabilities in comparison to their micro or bulk forms (McClements 2010).

The bioactives bioavailability are proportional to their bioaccessibility, *cellular* absorption and molecular transformation (Fig. 4.2). Thus, increasing either bioaccessibility or absorption as well as altering the molecular construction, which might occur during digestion, leads to a considerable increase in the bioavailability. The decrease of particle size and increase in the solubility of components leads to an increase of their surface area (surface/volume) and consequently intensifies their bioaccessibilities. The cellular absorption is affected by the components' pathway to reach the systematic flow and their metabolization place, where they might be utilized, stored or excreted (Holst and Williamson 2008; Huang et al. 2010). These processes define the components' concentrations and its metabolites at a particular site of the body, which depends on the time. The side effects or useful action of components depends on the components' concentration changes by time at the particular site-of-action. Accordingly, the measurement of the components' concentrations at a certain place of the body is essential to study its efficiency. It should be

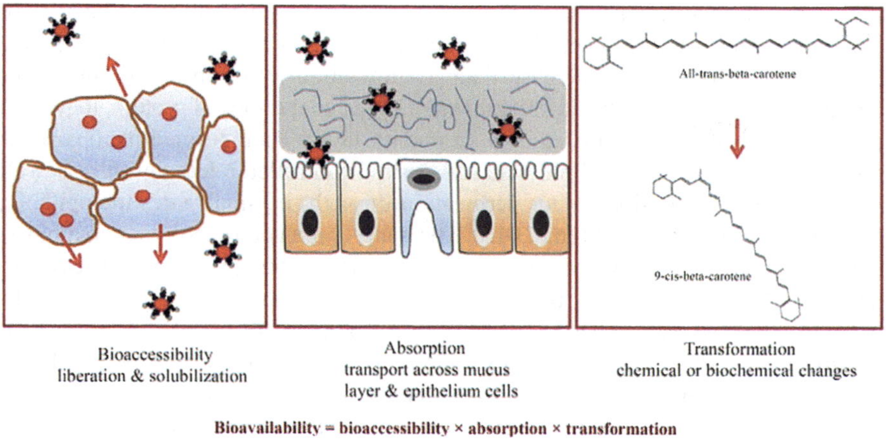

Fig. 4.2 Schematic delivery of bioactive components in food matrix (He and Hwang 2016)

noted that the components can be modified chemically or physically, before or after absorption, through passing into the gastrointestinal zone (Holst and Williamson 2008; McClements 2010). For instance, due to little water solubility of the CoQ10, it has very limited bioavailability. Thus, in order to improve the bioavailability of CoQ10, a lipid-free nano-CoQ10 system stabilized by various emulsifiers has been developed recently by Zhou et al. (2014), with mean particle sizes from 66.3 ± 1.5 nm to 92.7 ± 1.5 nm and stability of 180 days at 25 °C. Their bioavailability CoQ10 was also evaluated in rats. It was concluded that the produced nano-CoQ10 applies higher maximum plasma concentration than bulk equivalents of CoQ10 (Sekhon 2010; Zhou et al. 2014).

The nano-sized delivery systems can also control the interactions of components with biological medium and so their bio-distribution. Actually, the modification of the nano-delivery systems by grafting the hydrophilic molecules, such as polyethylene glycol, can be performed in order to control the molecular transformation and maximize their bioaccessibility and absorption (He and Hwang 2016).

The solid lipid nanoparticles (SLNs), the lipid nanoparticles with whole solid lipid matrix, have been proposed as the most efficient delivery systems for bioactive compounds that can either increase solubility, or improve the bioavailability and epithelium permeability. SLNs can control the physicochemical features of the nanoparticles for gastrointestinal penetration through either mucosal or lymphatic transport. In addition to nutraceuticals, the SLNs can nanoencapsulate various medicines in order to increase their oral bioavailability in cardiovascular-related diseases, diabetes, infection, cancers and osteoporosis treatments (Lin et al. 2017).

On the other hand, proteins have been reported as attractive bioavailability enhancer compounds for nutraceuticals due to their high abundance, desired biocompatibility and amphiphilic nature. Thus, the formulated nano-sized carriers with proteins as stabilizer or emulsifier can boost the bioavailability of incorporated compounds due to occurrence of proteins in their structures. Based on a noticeable rise in the cellular uptake of protein stabilized nanoparticles, the stabilization of anti-cancer compounds' nanoparticles, such as curcumin nanoparticles by proteins, are suggested in order to increase their accumulation into tumor cells in cancer therapy. To overcome the rapid breakdown of the nanoparticles of protein, they used the carboxymethylated chitosan as a secondary layer. Their formed nanoparticles presented a high functional lipid load and desired controlled release profile, as well as greater muco-adhesion, better permeation through the small intestine epithelia and higher cellular uptake (Teng 2015). The nano-laminated biopolymer coatings is another delivery system with the potential for encapsulating, protecting and controlling the release of functional lipid bioactives in food formulation uses. The thickness of layer, layers charge, composition, permeability and environmental parameters are some important factors affecting the absorption, release and digestion of active compounds. Thus, these delivery systems can either increase or decrease (control) the bioavailability of lipid compound in order to get the desired goal (McClements 2010).

Some inorganic nanoparticles can also be added into food formulations to improve their cellular uptake. For example, selenium nanoparticles are added into green tea product to yield increased selenium uptake, antioxidant activity and give a healthier product (Xu et al. 2003). Calcium, magnesium and iron nanoparticles can also be added to various food products as health supplements.

Tang et al. (2013) suggested a chitosan/polyglutamic acid nanocomposites for oral delivery of green tea catechins. Their gained nanoparticles showed retained antioxidant activity and increased para-cellular transport, despite passing throughout the gastrointestinal zone with harsh environment. Thus, their product can be used in various food formulations.

4.3.3 Control Release

Some bioactives often have side effects, including toxicity, despite their beneficial action such as fat-soluble vitamins or most pharmaceuticals. Thus, in order to provide effective delivery, the concentration control of the delivered bioactive compound over time and its accumulation on site of action should be carefully designed and controlled, which can be successfully performed by various nano-based delivery systems. These delivery systems are able to adjust the release of compounds because of their small size and high surface to volume ratios, capability to provide the intravascular distribution, steady release and stabilizing of bioactives. The nature and composition of delivery systems affects the release of inside bioactives to the environment. Thus, various types and concentrations of bioactive compound, stabilizing and emulsifying agents, nanoparticle size, can give various releasing property. It was shown that incorporating the metal or metal oxide nanoparticles with some bioactive compounds, especially the medicines, which are known as metallodrugs, can considerably affect their release in the aimed organ. For example, researchers have developed a metallodrug composed of zinc or vanadium for artificial insulin solutions (Sakurai et al. 2006), or platinum for cisplatin as anti-cancer (Jaouen and Metzler-Nolte 2010). Thus, the metal–organic compounds are more efficient than exactingly organic compounds for control release and gene delivery uses. The zeolite nanoparticles have also been claimed to have desired controlled release for sensor capsules or chemical, used in smart packaging (Dong et al. 2002). The responsive nano-sized delivery systems are more favorable, in which they would be capable to respond to an environment-changing signal, like pH, temperature, or ion concentrations etc. by swelling, dissolution, precipitation, collapsing, hydrophilic/hydrophobic balance (HLB) changing, degradation or other conforms (de Villiers et al. 2008; Kelland and Farrell 2000).

4.3.4 Antioxidant Nanoparticles

The favorable activity of antioxidant compounds, either hydrophilic or lipophilic compounds such as polyphenols, flavonoids and carotenoids, has been reported by previous researchers (Magalhães et al. 2008). Thus, supplementation of these compounds is required to neutralize the reactive oxygen and nitrogen substitutes and harmful radicals and prevent the oxidative stresses and treating various degenerative diseases and slowing down the aging process, as well. However, in spite of the *in vitro* antioxidant evaluations, all clinical analysis with these compounds have failed to prove their extensive benefits, in which the major restriction of dietary antioxidants is their very unstable chemical structure, naturally poor aqueous solubility and their deficiency in natural accumulation in target tissues (Gupta et al. 2015). For example, the anti-inflammatory, anti-hypertensive and antiallergic effects of quercetin has been shown through either *in vitro* or *in vivo* studies. Nevertheless, by now, it could not be used medicinally in formulations because of its less water solubility, chemical instability and thus low bioavailability. It was shown that the oral bioavailability of quercetin is just 1% in humans. Thus, nanoencapsulation of quercetin could either increase its *in vivo* antioxidant activity or extend its release and control the rate of first pass metabolism. Integrating of antioxidants into the nano-sized delivery systems can resolve their solubility, stability and drug release restrictions. Moreover, by conjugating the antioxidants into the hydrolytically biodegradable polymers' backbone, such as poly(β-amino esters), and formation of the pro-drug nanoparticles or nanogels can also enhance their activities and antioxidant efficiencies (Wang et al. 2016). It resulted that the encapsulation efficiency of antioxidant compounds is affected by the particle sizes of gained nanoparticles. Mostly, the loading of antioxidants in these systems show biocompatible equivalent and can decrease the oxidative stress of cells during long time periods (Gupta et al. 2015).

For example, ferulic acid and α-tocopherol loaded solid lipid nanoparticles (SLN) have been formulated by Oehlke et al. (2017). They concluded that while the antioxidant activity of α-tocopherol was increased significantly in their products, the activity of ferulic acid remained constant (Oehlke et al. 2017). Novel nanoantioxidants have also recently been developed using SiO_2–gallic acid nanoparticles.

These nanosystems showed considerable scavenging capacity against 2,2-diphenyl-1-picrylhydrazyl radicals (Deligiannakis et al. 2012). The health benefits of natural antioxidants can also be improved by their encapsulation in nanoemulsions. In a study, in oil-in-water sodium caseinate-stabilized nanoemulsions formulation, a stearin-rich milk fraction individually or in combination with α-tocopherol were used. They concluded that immobilized α-tocopherol in fat droplets provides considerable protection against degradation (Sekhon 2010).

Inorganic nanomaterials, particularly reformed fullerenes and cerium oxide nanoparticles, have also been found to have antioxidant activity and efficiently protect human cells from oxidative or reactive nitrogen damage. They can directly react with superoxide spices, since they act as superoxide dismutase mimetics *in vitro*. Finally, these nanomaterials have the potential to reduce the degradative radicals in biological organs (Karakoti et al. 2010).

Polyphenols, the secondary metabolites originated from plants, contain aromatic rings with hydroxyl groups, which provide a wide range of antioxidant compounds and, consequently, promote activity. Antimicrobial, anti-inflammatory, anti-mutagenic, anti-cancer and anti-carcinogenic properties are some other beneficial effects of these compounds. However, their little water-solubility, chemical instability and less bioavailability, which limited their uses in food applications, can be resolved by nanoencapsulation or carrying them into nano-sized delivery systems. It was resulted that the nanoencapsulation can offer a number of advantages, such as stability improvement, boosting the antioxidant activity, release adjustment and digestive protection etc. (Pimentel-Moral et al. 2018). Another example for antioxidant nanoparticles can be the preparation of the Pickering nanoemulsions stabilized by chitosan-tripolyphosphate in order to delivery of curcumin, by Shah et al. (2016). The Pickering nanoemulsions are the new category of nanoemulsions which have been formulated recently using solid particles as stabilizers as an alternative for emulsifiers. They reported amplified antioxidant ability for their resulted nanosized curcumin, than free curcumin (Shah et al. 2016).

4.3.5 Anti-browning Nanoparticles

Browning is an undesired reaction of foods, especially the fresh-cut fruits, that is caused by the changing of phenolic compounds into dark colored compounds in the presence of oxygen. Therefore, it should be controlled during either processing or preserving stages of the food to extend their shelf life. Besides the undesirable appearance, the browning can also be associated with annoying off-flavors and less nutritional value (Khan et al. 2016). It was shown that the nanoparticles can retard the browning reaction of fresh cut fruits, especially if it is associated with edible coatings. For example, the Fuji apple fresh cuts' shelf life was extended by coated nano sized-ZnO active packaging has been resulted previously. Moreover, the appearance of Fuji-apples fresh-cuts was preserved by decreasing their browning index (Li et al. 2011a). In another study, the quality and freshness of *Red Delicious* apple fresh cuts were preserved by packing them into poly-ε-caprolactone incorporated with nanocapsulated α-tocopherols (Zambrano-Zaragoza et al. 2014). It was also reported that the nano-packing containing silver nanoparticles could keep the browning rate of jujube at a minor level (Li et al. 2009). In other research, it was shown that silver nanoparticles can act as an anti-browning agent beside its antimicrobial and antioxidant activities. They prepared silver nanoparticles using inspired Longan fruit juice's phytochemicals as a reducing and stabilizing agent. Their presented silver nanoparticles could decrease the enzymatic browning of white cabbage, considerably. Their gained silver nanoparticles could control the enzymatic browning of cabbage by acting as either a reducing agent or free radical scavenger as well as changing the red-ox potential of the food (Khan et al. 2016).

4.3.6 Antimicrobial Nanoparticles

The microbial deterioration is the main concerning issue in the manufacturing, processing, handling and storage of food, since the microbial infections threaten human safety and lead to poor nutrition linked with dissuading foods. The nano-sized antimicrobial compounds have considerable influences on protection of food from deterioration and so prolonging their shelf life. Previous researchers have reported that the antibacterial activity of nanoparticles can be displayed as killing agents, growth inhibitors or antibiotic carriers (Dasgupta et al. 2015).

Various inorganic antimicrobials, such as metal and metal oxide nanomaterials as well as the organic ones such as essential oil nanoparticles, have been developed recently to propose effective antimicrobial compounds in food formulations. These compounds may cause the formation of reactive species and applying the oxidative stress to cells and, consequently, damage the microbial cells. Likewise, these compounds can also attach to the microbial cell walls or membranes and disrupt their structure and functions (He and Hwang 2016; Prakash et al. 2018).

Among organic antimicrobial nanoparticles, the essential oils are the most common components with antimicrobial activity. Various researchers demonstrated that size reduction of essential oils into nano-ranges could improve their antimicrobial activity, as well as solubility and stability. For instance, the antibacterial activity of chitosan—thyme essential oil nanoparticles and nanocapsules—have been studied by Sotelo-Boyás et al. (2017). They have prepared both nanoparticles and nanocapsules in less than 10 nm diameter and the results showed considerable inhibitory effects against both Staphylococcus aureus and Bacillus cereus (Sotelo-Boyás et al. 2017).

As stated before, the polymers are suitable stabilizers for either organic or metal and metal oxide nanoparticles. Low-density polyethylene, gelatin, isotactic polypropylene and polylactic acid are most widely used polymers in this regard. Chitosan, polyvinylpyrrolidone and polystyrene have also been used in formation of films that bind metal nanomaterials to deactivate the pathogens of food (He and Hwang 2016). Some of these stabilizing polymers can promote the antimicrobial activity of nanoparticles, due to its own antimicrobial effect. For example, chitosan is the most common biopolymer with antimicrobial activity, which can give considerable inhibitory effects against various microorganisms in combination to antimicrobial nanoparticles.

Most of the metal and metal oxide nanoparticles, which are classified as inorganic nanoparticles, show antimicrobial activity. Silver nanoparticles are the most common used nano-sized antimicrobial in the food industry. Depending on the size reduction route of silver, the nanoparticles can be stabilized by either organic or inorganic stabilizers. In preparation of inorganic nanoparticles through chemical process, the synthetic or natural reductants are also needed. The influence of silver nanoparticles on food components in wheat have been studied by Nawrocka and Cieœla (2013). They showed that the microbial contamination of the covered grains with nano-silver have been decreased, significantly, during storage time. They haven't observed any significant changes on falling numbers and gluten parameters of covered grains with sodium citrate stabilized silver nanoparticles. However, the moisture content, and secondary structure of grain proteins have been changed, considerably (Nawrocka and Cieœla 2013).

TiO_2 nanoparticles can also stop the microbial growth in food, due to their photocatalyst activity, which can endorse peroxidation of the phospholipids in cell membranes of microorganisms (Dasgupta et al. 2015). TiO_2-coated wrapping film can diminish the *E. coli* contamination of food surfaces and fecal coliform of water, significantly (Martirosyan and Schneider 2014).

The nano-antimicrobial compounds can be used in food packaging and edible coatings as well as food formulations. However, in incorporation of antimicrobial nanoparticles into a food packaging matrix, their little and negligible migrations from the packaging into the food are desired. The migration of these compounds into a food matrix depends on the particle characteristics, stabilizer type, polymer nature, food structure and the food production, packaging and storage conditions such as humidity, temperature, light exposure and storage period. Due to safety concerns of especially inorganic nanomaterials, their limited loading into food matrices are anticipated (Dasgupta et al. 2015; Martirosyan and Schneider 2014).

The antibacterial action of carbon nanotubes was also reported, recently, due to its ability in rupturing the microbial cells and initiating their permanent damages. However, due to their cytotoxicity against human cells, it can be used as just an antibacterial additive in packaging formulations (Dasgupta et al. 2015). Most of the inorganic antimicrobial nanoparticles offer greater temperature stabilities and low volatility as compared to organic ones. Thus, they would be preferred in foods with more severe processing conditions. Nanomaterials offer additional stability and provide sustaining antimicrobial action, due to possible control of their releases into the food system (Dasgupta et al. 2015; Tsagkaris et al. 2018).

4.3.7 Mechanical Strength

Nano-clay are the most common nanoparticles which were added to packaging polymers in order to improve their gas barrierity and mechanical strength. Nano-titanium dioxide and nano-titanium nitride can be used as ultraviolet (UV) protecting and mechanical reinforcement and processing aiding agents, respectively. Recently, the gelatin-agar films holding TiO_2 nanoparticles have been presented in order to preserve the light-sensitive foods. It was shown that by increasing the TiO_2 content of the films, the water vapor permeability and swelling ratio has been decreased and increased, respectively, almost certainly due to the hydrophilic nature of the TiO_2 nanoparticles. In addition, the gas barrier of films have been improved and opacity was increased by TiO_2 concentration (Vejdan et al. 2016).

Recently, functionalized silver nanoparticles on Cu-doped ZnO composites were embedded into poly lactic acid (PLA) matrix for food packaging uses. The thermal and UV irradiation stability, tensile strength, transmittance, gas and water vapor barrier of PLA could be improved significantly. Moreover, the modified polymer showed considerable antimicrobial activity (Vasile et al. 2017).

The performance of biodegradable and environmentally friendly biopolymer-based packaging can also be improved by the incorporation of nanoparticles into their formulations. The nanoparticles can improve the limiting drawbacks of biopolymers in food packaging uses such as their high permeability to gases and water vapor, poor mechanical strength and less heat resistance etc. For example, the incorporation of montmorillonite (less than 5%) into poly lactic acid could increase its tensile modulus and yield strength as well as decrease its oxygen permeability. Incorporation of nano-clay in starch polymers could also improve their thermal stability, moisture barrier and mechanical properties (Othman 2014). Figure 4.3 shows the nanoparticles that can be added to food packaging formulations in order to strengthen their structures.

Edible coatings or films are biodegradable barriers of oxygen, water and lights, applied onto the foods in order to extend their shelf life. Nanotechnology provides the development of edible coating/films with nano-structured components, which

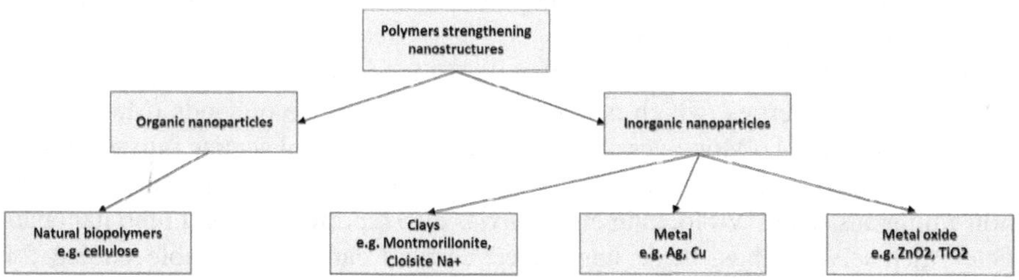

Fig. 4.3 The classification of nanostructure's natures used for strengthening of food packaging polymers

are recognized as edible nano-coatings/films. These nano-coatings/films can be used for all kinds of foods such as vegetables, meats, bakery products and cheese etc. They can also deliver antimicrobials, flavors, colors, antioxidants, anti-browning and enzymes to the coated food, leading to their shelf life extension as well as sensory characteristics enhancement, even after opening its packaging (Sekhon 2010). Metal or metal oxide nanoparticles can be used in edible coating/film formulations in order to improve their characteristics such as gas barrierity, anti-radiation and antimicrobial activities. For instance, protein based edible films conjugated with TiO_2 are recently suggested by researchers (Li et al. 2011b). They reported better adhesion of edible coatings on fruits, decreased UV absorption and promoted tensile strength and fluorescence films with antimicrobial activity, depending on TiO_2 concentration.

4.3.8 Absorbent

Amorphous silica has also been used for clarifying beers and wine, removing unwanted components such as the bitter taste in some plant products. Silver nanoparticles can also absorb and break down the ethylene, which contributes to extending the shelf life of vegetables and fruits. It was also reported that the calcium silicate nanoparticles can adsorb the Ag ions from a system. Thus, it can remove the penetrated Ag ions from antibacterial packaging incorporated with silver nanoparticles into the food system (Dasgupta et al. 2015; Tsagkaris et al. 2018).

4.3.9 UV-Blocker Nanoparticles

UV light possess greater energy than visible light, and thus, it can breakdown the chemical bonds of compounds. Several foods are exposed to sunlight, or artificial fluorescent lights, during process, storing and transport etc. Both the sun and fluorescent lights produce UV and visible light. Some food components, such as colorants, lipids, proteins and vitamins, can react with light and produce undesired compounds. Thus, the food packaging with UV-blocking activity are favored, mostly (Duncan and Hannah 2012).

As stated before, it was concluded that the TiO_2 nano-scales TiO_2 express greater UV-blocking properties, which is beneficial for the storage of foods (Martirosyan and Schneider 2014). Moreover, the incorporated tocopherol acetate into solid lipid nanoparticles showed a higher UV-blocking capacity than their equivalent convention emulsions. The UV-blocking efficacy was also reported for solid lipid nanoparticles themselves. Thus, these nano-sized carriers can be a suitable choice for delivery of UV-blocking compounds (Wissing and Müller 2001).

4.3.10 Pigmentation

The food colorant additives in food systems should be used only in obedience with the approved uses, specifications and restrictions by the FDA. Thus, there is limited applicable colorant compounds for food formulation. Since the color of the materials is a surface dependent characteristic, nanotechnology provides the color adjustment of these compounds. Thus, a various range of nano-sized color additives have been evaluated and manufactured. The colorant nanomaterials can also have either inorganic or organic nature. TiO_2 nanoparticles is an inorganic, FDA approved food colorant compound, in less than 1% w/w concentration uses. The mixture of TiO_2 with AL_2O_3 or SiO_2, in less than 2% w/w, are also accepted for food formulations. Candies, sweets and chewing gums can be formulated with TiO_2 nanoparticles as a colorant ingredient (He and Hwang 2016; Martirosyan and Schneider 2014).

While the metal and metal oxide pigment nanoparticles are mostly used in food packaging formulations in order to provide market desirability, the organic pigment nanoparticles were added to food systems for either pigmentation or health promoting targets (He and Hwang 2016; Huang et al. 2010; Martins et al. 2016).

Anthocyanins, originated from various fruits and vegetables, can be used as either natural colorants or antioxidants. The uses of anthocyanins are preferred due to their intense and bright attractive colors and good water-solubility. However, their less bioavailability and restricted chemical stability against environmental conditions, such as light, oxygen, heat, pH, enzymes and occurrence of ascorbic acid etc. limits their commercial uses in various food formulations. Nanoencapsulation of them can improve their stabilities considerably (Arroyo-Maya and McClements 2015; He et al. 2017). Table 4.2 shows some nanotechnology offered techniques for increasing the stability and bioavailability of anthocyanins.

Carotenoids, extracted from either plants or microbial cells, are also another set of natural colorants, which have various applications in food products' pigmentation. Like anthocyanins, they also have various health promoting benefits for human cells. Their antioxidant activity has been scientifically confirmed by various researchers. However, they are almost water-insoluble, unlike the anthocyanins. Thus, incorporation of these colorants is challenging in food and beverage formulations. Moreover, the chemical structure of carotenoids are susceptible to heat, light and oxygen of the storage environment. The studies also reported limited bioavailability and cellular uptake for these compounds, as well. The nano-based delivery systems incorporated with carotenoids, can resolve most of the mentioned problems and increase their water solubilities, bioavailabilities, and stabilities (Anarjan and Tan 2013a; Martins et al. 2016; Tan and Nakajima 2005). Some of the research, which has been completed in this regard, is shown in Table 4.3.

The nanoencapsulation of phenolic compounds, curcumin, annatto, bixin and norbixin in order to increase their stabilities, water solubility and bioavailability have also been studied recently by several research groups. In some research the

Table 4.2 The nanotechnology offered techniques for improving the stability and bioavailability of anthocyanins

Pigment	Delivery system	Encapsulation, stabilization and emulsifying compounds	Description	References
Anthocyanin (originated from Açaí berry)	Anthocyanins-loaded nanoparticles	EUDRAGIT® L100, polyethylene glycol 2000 (PEG 2000), polysorbate 80	Increases bioavailability	Danay et al. (2017)
Anthocyanin	Electrostatic complexation	Whey protein isolate (WPI) Beet pectin (BP)	D < 200 nm Increases thermal stability, decreases antioxidant activity, have similar color intensity	Arroyo-Maya and McClements (2015)
Anthocyanin	Anthocyanins loaded nanoparticles	Carboxymethyl chitosan (CMC), chitosan hydrochloride	D = 214.83 nm Increases gastrointestinal stability, used in model beverage system	He et al. (2017)
Anthocyanin	Complexation (intermolecular stacking between chondroitin sulfate and Anthocyanin)	Chondroitin sulfate	D = 300 nm Increases pH stability, the system has a potential as a carrier for antioxidants or water-soluble drugs.	Jeong and Na (2012)
Anthocyanin	Complexation	Native whey protein; denatured whey protein; citrus pectin; and beet pectin (these biopolymers were used separately in complexation)	Increases the stability in model beverage systems for longer storage periods, heat denatured whey protein produced more stable product	Chung et al. (2015)

effects of nano-sized colorants at various food systems, such as beverages and desserts etc., have also been investigated. Their results indicated that the nano-sized natural food colorants can offer more intense colors, more homogeneous systems and greater cellular uptake and, consequently, larger health promoting effects in incorporated foods (Amjadi et al. 2018; Anarjan and Tan 2013a; Ghayour et al. 2019; Wang et al. 2018).

Although some metal nanoparticles can stabilize the organic pigments, others would accelerate their degradation rates. Thus, in pigmentation of food products, the interaction of the colorant components should be considered (Martins et al. 2016).

Table 4.3 Some offered nano-based delivery systems for carotenoids for pigmentation uses

Pigment	Delivery system	Encapsulation, stabilization and emulsifying compounds	Description	References
β-carotene (and mixture of β-carotene, α-carotene and lutein)	Lipid core nanoparticles	Poly-ε-caprolactone sorbitan monostearate Triglycerides of the capric and caprylic acids polysorbate 80	D < 191 nm The β-carotene retention in the mixed carotenoids lipid core nanoparticles were more than β-carotene ones The color intensities and stabilities of products remained constant over more than 100 days storage	da Silva et al. (2016)
β-carotene	Solid lipid nanoparticles and liquid lipid nanoparticles	Cocoa butter and/ or hydrogenated palm oil Tween 80	D < 200 nm While the chemical and physical stabilities were increased for both systems, the liquid nanoparticles showed better efficiency in nanoencapsulation of the β-carotene	Qian et al. (2013)
β-carotene	Nanodispersion	Modified OSA-starch	D = 200–300 nm Gastrointestinal stability of β-carotene was increased Used in model beverage system	de Paz et al. (2012)
β-carotene	Nanodispersion	Tween 20	D = 9–280 nm The stability and color intensity of β-carotene were increased	Silva et al. (2011)
Lutein	Nanodispersion	Tween 80, short chain triglycerides/ or medium chain triglycerides/ or long chain triglycerides	D = 200 nm Nanoemulsions could successfully encapsulated and protected the lutein	Surh et al. (2017)
Lutein	Nanoemulsions	Whey protein isolate	D = 68 nm The cellular uptake of lutein in nanoemulsions was increased considerably	Teo et al. (2017)

(continued)

Table 4.3 (continued)

Pigment	Delivery system	Encapsulation, stabilization and emulsifying compounds	Description	References
Lutein	Nanodispersions	Tween 80 and sodium dodecyl sulfate sodium caseinate (individually or in combinations)	D = 66.20–125.25 nm The cellular uptake and color intensities were changed based on selected emulsifier	Tan et al. (2016a, b)
Astaxanthin	Nanodispersions	Tween 20, sodium caseinate, Arabic gum (individually or in combinations)	D = 98–157 nm The color intensity of samples were changed based on used stabilizers. The three-component stabilizer could produce the most desired nanodispersions The cellular uptake of astaxanthin in prepared nanodispersions were increased considerably	Anarjan et al. (2012)
Zeaxanthin	Nanoparticles Nanoemulsions	Chia seed oil Cactus mucilage Tween 80	D = 184 nm The cactus cladode mucilage could protect zeaxanthin from structural degradation, significantly	de Campo et al. (2018)
Lycopene	Lipid core nanocapsules	Polymer (PCL), caprylic/capric triglycerides, sorbitan monostearate	D = 193 nm The physical stability was improved However, the color increases for product happened during storage	dos Santos et al. (2015)
Lycopene	Nanodispersions	Tween 20	Minimum D = 73 nm The physical stability of lycopene nanodispersions was increased. However, their lycopene content was decreased by a first order kinetic model	Anarjan and Jouyban (2017)

The simultaneous extraction of natural pigmented from food manufacturing by-products, like tomato or grape skins, and their size reduction into nano-ranges, can also be efficient and economical for food-colorant uses (Anarjan and Jouyban 2017).

Biotechnology proposed the effective mass production of pigments, such as carotenoids, anthocyanins and betalains, using gene manipulation, *in vitro* culturing and fermentation of various plant, animal and microbial cells. On the other side, the nanotechnology provides their delivery into the food systems used nano-sized carriers in order to increase their stabilities, solubility and cellular uptake. Thus, nano-biotechnology presents considerable achievements in food pigmentation issues (Martins et al. 2016; Nickols-Richardson 2017).

4.3.11 Flavor

One of most vital parts of the food system is flavor, which supplies sensory insights of smell and taste to boost the total eating experience. Nanoencapsulation can increase flavor retention and release, considerably, in food. For instance, SiO_2 nanomaterials can deliver the flavors or fragrances in food products (He and Hwang 2016). Nanoencapsulated flavors as well as flavor enhancers can be added to cooking oils to boost the sensory characteristics of cooked food by gradually binding or removing some of the compounds from them (Sozer and Kokini 2009). The nanoencapsulation can mask unpleasant fishy odors of oils as well (Chen et al. 2016). Thus, the nanoencapsulation of food flavors can give better control of liquid flavors by converting them into a dry form, increasing stability in exposure to heat, light and oxygen, enhance shelf life, control release of volatile flavor components and masking off-odors and off-tastes. Coacervation, inclusion, nanoprecipitation, liposomes, nanoemulsions and solid–lipid nanoparticles are some nanoencapsulation routes for food flavors (Farhath et al. 2018; Prakash et al. 2018).

4.3.12 Nanosensors/Nanobiosensors

Nanosensors are included in food packaging to sense the pathogens or chemical contaminants during food spoilages, providing accurate expiration dates and real-time rank of food freshness. They are also able to respond to the oxygen exposure, temperature and humidity changes of the environment during storage.

Time–temperature indicators in smart packaging are designed to sense, record and interpret the safety of packaged food, especially if it is stored in non-optimal conditions. For example, recently, smart packaging, based on gold nanoparticles, has established a system for chilled foods.

The microorganism-based food spoilage may be detected by several gas sensors as well. Metal oxide nanoparticles are efficient for detecting various gases. The conducting polymer nanoparticles can also detect microorganisms due to their different gas emission activities. TiO_2 nanoparticle-based biosensors have been developed for O_2 detection, by photosensitizing the methylene blue reduction by triethanolamine in polymer matrix using UV light. A rapid and precise detection of *E. coli* have also been possible by a proposed nanotechnology-based nanosensor which uses a bacterium protein, a nano silicon chip and a digital camera in which the protein on the chip binds with any *E. coli* and changes the surface of the chip. These changes can be detected by an incorporated digital camera. The attendance, type and concentration of contaminant microbes can also be detected based on nanowire stabilized antigen and antibodies, as the pathogen particular antibodies are attached to the placed nanowires in the food (Dasgupta et al. 2015). Fluorescent nanoparticles for in situ detection of pathogen in ground beef using antibody-conjugated silica has also been developed. Luciferase nanosensors can also detect specific pathogen by recording the luminescence dye due to their attachment onto certain bacteria such as *Salmonella* and *Campylobacter*. Magnetic nanoparticles in combination with adenosine triphosphate bioluminescence have been proposed recently to detect the main food spoilage microorganisms (Augustin and Sanguansri 2009; Bajpai et al. 2018). The nanoencapsulated enzymes, antibodies, antigens, polysaccharides and glycol-proteins can be suggested as efficient nanobiosensors for sensing special compounds in foods, which are indicators for various microbial or chemical spoilage of foods (Bajpai et al. 2018; He and Hwang 2016; Martirosyan and Schneider 2014).

4.3.13 Other Active Nanoparticles

The metal or metal oxide nanoparticles can also be used as anti-odorant in various food packaging. For example, the researchers reported anti-odorant activity, for silver nanoparticles, beside their antimicrobial action. ZnO nanoparticles were also used in essential oil embedded antimicrobial packaging films in order to remove any odors of volatiles. The porous nanomaterials can be effectively be used as anti-odors due to its great surface area and, consequently, its considerable ability for adsorption of various volatile compounds.

Nanofluids, a suspension of nanoparticles in fluids with enhanced thermophysical properties such as thermal conductivity and viscosity, can also be used in order to provide enhanced and controllable heat transfer for various food processes. Mineral oil-based nanofluids containing silver nanoparticles prepared through a one-step physical vapor condensation method is an efficient nanofluid for food processing uses such as sterilization and evaporation etc.

The anti-caking agent, which is added to powder food formulations in order to prevent the lump formation, can also be sized reduced into nano-ranges. Their mechanism of actions are either absorbing excess moisture or making them hydrophobic by proper coating of the particles. Calcium silicate is the most common anti-caking agent either in food uses, such as table sugars, spices and flavors, or non-food areas such as fertilizers, detergents and cosmetic formulations. SiO_2 with E number of 551 (E551) is mostly added to pastes as an anti-caking agent to thicken and maintain their flow properties (He and Hwang 2016). TiO_2 (E171) is also added to granular and powdered foods as anti-caking agents (Martirosyan and Schneider 2014).

Probiotics are mixes of alive essential and useful bacterial species that can help the digestion system and support human health. Thus, they can be integrated in various food formulations such as desserts and drinks. Encapsulation of probiotics can extend their shelf life and successfully deliver them to definite gastro-intestinal zones, where they interact with particular receptors. The encapsulating material should provide the nutrition requirement of the encapsulated bacterial species. Otherwise, the nutrient materials should also be encapsulated together with the alive cells. Biopolymers stabilized by non-covalent forces have been revealed to provide substantial progress in this regard (Sekhon 2010). The nanoencapsulation of necessary nutrients for probiotics and their addition into food formulations can also provide the optimum conditions for incorporated probiotics growth (Augustin and Sanguansri 2009; Dasgupta et al. 2015).

The self-cleaning nanostructured coatings with dirt-repellent surfaces can also be applied onto food packaging in order to increase packed food safety by inhibiting the microorganism growth.

References

Amjadi S, Ghorbani M, Hamishehkar H, Roufegarinejad L. Improvement in the stability of betanin by liposomal nanocarriers: its application in gummy candy as a food model. Food Chem. 2018;256:156–62.

Anarjan N, Jouyban A. Preparation of lycopene nanodispersions from tomato processing waste: effects of organic phase composition. Food Bioprod Process. 2017;103:104–13.

Anarjan N, Tan CP. Chemical stability of astaxanthin nanodispersions in orange juice and skimmed milk as model food systems. Food Chem. 2013a;139(1):527–31.

Anarjan N, Tan CP. Developing a three component stabilizer system for producing astaxanthin nanodispersions. Food Hydrocoll. 2013b;30(1):437–47.

Anarjan N, Mirhosseini H, Baharin BS, Tan CP. Effect of processing conditions on physicochemical properties of astaxanthin nanodispersions. Food Chem. 2010;123(2):477–83.

Anarjan N, Mirhosseini H, Baharin BS, Tan CP. Effect of processing conditions on physicochemical properties of sodium caseinate-stabilized astaxanthin nanodispersions. LWT-Food Sci Technol. 2011a;44(7):1658–65.

Anarjan N, Tan CP, Ling TC, Lye KL, Malmiri HJ, Nehdi IA, Cheah YK, Mirhosseini H, Baharin BS. Effect of organic-phase solvents on physicochemical properties and cellular uptake of astaxanthin nanodispersions. J Agric Food Chem. 2011b;59(16):8733–41.

Anarjan N, Tan CP, Nehdi IA, Ling TC. Colloidal astaxanthin: preparation, characterisation and bioavailability evaluation. Food Chem. 2012;135(3):1303–9.

Anarjan N, Jafarizadeh Malmiri H, Ling TC, Tan CP. Effects of pH, ions, and thermal treatments on physical stability of astaxanthin nanodispersions. Int J Food Prop. 2014;17(4):937–47.

Anarjan N, Fahimdanesh M, Jafarizadeh-Malmiri H. β-Carotene nanodispersions synthesis by three-component stabilizer system using mixture design. J Food Sci Technol. 2017;54(11):3731–6.

Arroyo-Maya IJ, McClements DJ. Biopolymer nanoparticles as potential delivery systems for anthocyanins: fabrication and properties. Food Res Int. 2015;69:1–8.

Augustin MA, Sanguansri P. Nanostructured materials in the food industry. Adv Food Nutr Res. 2009;58:183–213.

Augustin MA, Sanguansri L, Bode O. Maillard reaction products as encapsulants for fish oil powders. J Food Sci. 2006;71(2):E25–32.

Bajpai VK, Kamle M, Shukla S, Mahato DK, Chandra P, Hwang SK, Kumar P, Huh YS, Han Y-K. Prospects of using nanotechnology for food preservation, safety, and security. J Food Drug Anal. 2018;26(4):1201–14.

Chaudhry Q, Scotter M, Blackburn J, Ross B, Boxall A, Castle L, Aitken R, Watkins R. Applications and implications of nanotechnologies for the food sector. Food Addit Contam Part A Chem Anal Control Expo Risk Assess. 2008;25(3):241–58.

Chen XW, Chen YJ, Wang JM, Guo J, Yin SW, Yang XQ. Phytosterol structured algae oil nanoemulsions and powders: improving antioxidant and flavor properties. Food Funct. 2016;7(9):3694–702.
Chen F, Fan GQ, Zhang Z, Zhang R, Deng ZY, McClements DJ. Encapsulation of omega-3 fatty acids in nanoemulsions and microgels: impact of delivery system type and protein addition on gastrointestinal fate. Food Res Int. 2017;100:387–95.
Cheong JN, Tan CP, Man YBC, Misran M. α-Tocopherol nanodispersions: preparation, characterization and stability evaluation. J Food Eng. 2008;89(2):204–9.
Chung C, Rojanasasithara T, Mutilangi W, McClements DJ. Enhanced stability of anthocyanin-based color in model beverage systems through whey protein isolate complexation. Food Res Int. 2015;76:761–8.
da Silva MM, Nora L, Cantillano RFF, Paese K, Guterres SS, Pohlmann AR, Costa TMH, Rios AO. The production, characterization, and the stability of carotenoids loaded in lipid-core nanocapsules. Food Bioprocess Technol. 2016;9(7):1148–58.
Danay D, Monique K, Jacques R, José CC. Formulation and characterization of anthocyanins-loaded nanoparticles. Curr Drug Deliv. 2017;14(1):54–64.
Dasgupta N, Ranjan S, Mundekkad D, Ramalingam C, Shanker R, Kumar A. Nanotechnology in agro-food: from field to plate. Food Res Int. 2015;69:381–400.
de Campo C, Dick M, Pereira dos Santos P, Haas Costa TM, Paese K, Stanisçuaski Guterres S, de Oliveira Rios A, Hickmann Flôres S. Zeaxanthin nanoencapsulation with Opuntia monacantha mucilage as structuring material: characterization and stability evaluation under different temperatures. Colloids Surf A Physicochem Eng Asp. 2018;558:410–21.
de Paz E, Martín T, Estrella A, Rodríguez-Rojo S, Matias AA, Duarte CMM, Cocero MJ. Formulation of β-carotene by precipitation from pressurized ethyl acetate-on-water emulsions for application as natural colorant. Food Hydrocoll. 2012;26(1):17–27.
de Souza Simões L, Madalena DA, Pinheiro AC, Teixeira JA, Vicente AA, Ramos ÓL. Micro- and nano bio-based delivery systems for food applications: In vitro behavior. Adv Colloid Interf Sci. 2017;243:23–45.
de Villiers MM, Aramwit P, Kwon GS. Nanotechnology in drug delivery. New York: Springer; 2008.
Deligiannakis Y, Sotiriou GA, Pratsinis SE. Antioxidant and antiradical SiO_2 nanoparticles covalently functionalized with gallic acid. ACS Appl Mater Interfaces. 2012;4(12):6609–17.
Dong A, Wang Y, Tang Y, Ren N, Zhang Y, Gao Z. Hollow zeolite capsules: a novel approach for fabrication and guest encapsulation. Chem Mater. 2002;14(8):3217–9.
dos Santos PP, Paese K, Guterres SS, Pohlmann AR, Costa TH, Jablonski A, Flôres SH, Rios AO. Development of lycopene-loaded lipid-core nanocapsules: physicochemical characterization and stability study. J Nanopart Res. 2015; 17(2).
Duncan SE, Hannah S. 15—Light-protective packaging materials for foods and beverages. In: Yam KL, Lee DS, editors. Emerging food packaging technologies. Cambridge: Woodhead Publishing; 2012. p. 303–22.
Esquerdo VM, Silva PP, Dotto GL, Pinto LAA. Nanoemulsions from unsaturated fatty acids concentrates of carp oil using chitosan, gelatin, and their blends as wall materials. Eur J Lipid Sci Technol. 2018;120(2):1700240.
Farhath K, Syeda Juveriya F, Ilaiyaraja N, Anand T, Patil MM, Singsit D, Sharma GK. Nanoencapsulation of flavors. In: Selvamuthukumaran M, Pathak Y, editors. Flavors for nutraceutical and functional foods. Boca Raton, FL: CRC Press; 2018. p. 274–312.
Ghayour N, Hosseini SMH, Eskandari MH, Esteghlal S, Nekoei AR, Hashemi Gahruie H, Tatar M, Naghibalhossaini F. Nanoencapsulation of quercetin and curcumin in casein-based delivery systems. Food Hydrocoll. 2019;87:394–403.
Golfomitsou I, Mitsou E, Xenakis A, Papadimitriou V. Development of food grade O/W nanoemulsions as carriers of vitamin D for the fortification of emulsion based food matrices: a structural and activity study. J Mol Liq. 2018;268:734–42.

Gunasekaran T, Haile T, Nigusse T, Dhanaraju MD. Nanotechnology: an effective tool for enhancing bioavailability and bioactivity of phytomedicine. Asian Pac J Trop Biomed. 2014;4(Suppl 1):S1–7.

Gupta P, Authimoolam SP, Hilt JZ, Dziubla TD. Quercetin conjugated poly (β-Amino Esters) nanogels for the treatment of cellular oxidative stress. Acta Biomater. 2015;27:194–204.

He X, Hwang H-M. Nanotechnology in food science: functionality, applicability, and safety assessment. J Food Drug Anal. 2016;24(4):671–81.

He B, Ge J, Yue P, Yue X, Fu R, Liang J, Gao X. Loading of anthocyanins on chitosan nanoparticles influences anthocyanin degradation in gastrointestinal fluids and stability in a beverage. Food Chem. 2017;221:1671–7.

Holst B, Williamson G. Nutrients and phytochemicals: from bioavailability to bioefficacy beyond antioxidants. Curr Opin Biotechnol. 2008;19(2):73–82.

Hou X, Lei S, Qiu S, Guo L, Yi S, Liu W. A multi-residue method for the determination of pesticides in tea using multi-walled carbon nanotubes as a dispersive solid phase extraction absorbent.Food Chem. 2014;153:121–9.

Huang Q, Yu H, Ru Q. Bioavailability and delivery of nutraceuticals using nanotechnology. J Food Sci. 2010;75(1):R50–7.

Jaouen G, Metzler-Nolte N. Medicinal organometallic chemistry. Berlin: Springer; 2010.

Jemaa MB, Falleh H, Serairi R, Neves MA, Snoussi M, Isoda H, Nakajima M, Ksouri R. Nanoencapsulated Thymus capitatus essential oil as natural preservative. Innov Food Sci Emerg Technol. 2018;45:92–7.

Jeong D, Na K. Chondroitin sulfate based nanocomplex for enhancing the stability and activity of anthocyanin. Carbohydr Polym. 2012;90(1):507–15.

Kadappan AS, Guo C, Gumus CE, Bessey A, Wood RJ, McClements DJ, Liu Z. The efficacy of nanoemulsion-based delivery to improve vitamin D absorption: comparison of in vitro and in vivo studies. Mol Nutr Food Res. 2018; 62(4).

Karakoti A, Singh S, Dowding JM, Seal S, Self WT. Redox-active radical scavenging nanomaterials. Chem Soc Rev. 2010;39(11):4422–32.

Karthik P, Anandharamakrishnan C. Enhancing omega-3 fatty acids nanoemulsion stability and in-vitro digestibility through emulsifiers. J Food Eng. 2016;187:92–105.

Katouzian I, Jafari SM. Nano-encapsulation as a promising approach for targeted delivery and controlled release of vitamins. Trends Food Sci Technol. 2016;53:34–48.

Kelland LR, Farrell NP. Platinum-based drugs in cancer therapy. New York: Humana Press; 2000.

Kesarwani K, Gupta R. Bioavailability enhancers of herbal origin: an overview. Asian Pac J Trop Biomed. 2013;3(4):253–66.

Khan AU, Wei Y, Ahmad A, Haq Khan ZU, Tahir K, Khan SU, Muhammad N, Khan FU, Yuan Q. Enzymatic browning reduction in white cabbage, potent antibacterial and antioxidant activities of biogenic silver nanoparticles. J Mol Liq. 2016;215:39–46.

Li H, Li F, Wang L, Sheng J, Xin Z, Zhao L, Xiao H, Zheng Y, Hu Q. Effect of nano-packing on preservation quality of Chinese jujube (Ziziphus jujuba Mill. var. inermis (Bunge) Rehd). Food Chem. 2009;114(2):547–52.

Li X, Li W, Jiang Y, Ding Y, Yun J, Tang Y, Zhang P. Effect of nano-ZnO-coated active packaging on quality of fresh-cut 'Fuji' apple. Int J Food Sci Technol. 2011a;46(9):1947–55.

Li Y, Jiang Y, Liu F, Ren F, Zhao G, Leng X. Fabrication and characterization of TiO_2/whey protein isolate nanocomposite film. Food Hydrocoll. 2011b;25(5):1098–104.

Lin C-H, Chen C-H, Lin Z-C, Fang J-Y. Recent advances in oral delivery of drugs and bioactive natural products using solid lipid nanoparticles as the carriers. J Food Drug Anal. 2017;25(2):219–34.

Magalhães LM, Segundo MA, Reis S, Lima JLFC. Methodological aspects about in vitro evaluation of antioxidant properties. Anal Chim Acta. 2008;613(1):1–19.

Martins N, Roriz CL, Morales P, Barros L, Ferreira ICFR. Food colorants: challenges, opportunities and current desires of agro-industries to ensure consumer expectations and regulatory practices. Trends Food Sci Technol. 2016;52:1–15.

Martirosyan A, Schneider Y-J. Engineered nanomaterials in food: implications for food safety and consumer health. Int J Environ Res Public Health. 2014;11(6):5720–50.

Mayer S, Weiss J, McClements DJ. Vitamin E-enriched nanoemulsions formed by emulsion phase inversion: factors influencing droplet size and stability. J Colloid Interface Sci. 2013;402:122–30.

McClements DJ. Design of nano-laminated coatings to control bioavailability of lipophilic food components. J Food Sci. 2010;75(1):R30–42.

Nawrocka A, Cieœla J. Influence of silver nanoparticles on food components in wheat. Int Agrophys. 2013;27:49–55.

Nickols-Richardson SM. Nanotechnology: implications for food and nutrition professionals. J Am Diet Assoc. 2017;107(9):1494–7.

Oehlke K, Behsnilian D, Mayer-Miebach E, Weidler PG, Greiner R. Edible solid lipid nanoparticles (SLN) as carrier system for antioxidants of different lipophilicity. PLoS One. 2017;12(2):171662.

Osorio-Tobón JF, Silva EK, Meireles MAA. 3—Nanoencapsulation of flavors and aromas by emerging technologies. In: Grumezescu AM, editor. Encapsulations. London: Academic; 2016. p. 89–126.

Othman SH. Bio-nanocomposite materials for food packaging applications: types of biopolymer and nano-sized filler. Agric Agric Sci Procedia. 2014;2:296–303.

Ozturk B, Argin S, Ozilgen M, McClements DJ. Formation and stabilization of nanoemulsion-based vitamin E delivery systems using natural biopolymers: whey protein isolate and gum arabic. Food Chem. 2015a;188:256–63.

Ozturk B, Argin S, Ozilgen M, McClements DJ. Nanoemulsion delivery systems for oil-soluble vitamins: influence of carrier oil type on lipid digestion and vitamin D3 bioaccessibility. Food Chem. 2015b;187:499–506.

Palza H. Antimicrobial polymers with metal nanoparticles. Int J Mol Sci. 2015;16(1):2099–116.

Park BK, Kim D, Jeong S, Moon J, Kim JS. Direct writing of copper conductive patterns by ink-jet printing. Thin Solid Films. 2007;515(19):7706–11.

Paszkiewicz M, Tyma M, Jakubus A, Stepnowski P. Recent applications of carbon nanotubes as sorbents for the extraction of pharmaceutical residues. Curr Anal Chem. 2016;12(4):268–79.

Pimentel-Moral S, Teixeira MC, Fernandes AR, Arráez-Román D, Martínez-Férez A, Segura-Carretero A, Souto EB. Lipid nanocarriers for the loading of polyphenols—a comprehensive review. Adv Colloid Interf Sci. 2018;260:85–94.

Prakash B, Kujur A, Yadav A, Kumar A, Singh PP, Dubey NK. Nanoencapsulation: an efficient technology to boost the antimicrobial potential of plant essential oils in food system. Food Control. 2018;89:1–11.

Qian C, Decker EA, Xiao H, McClements DJ. Impact of lipid nanoparticle physical state on particle aggregation and β-carotene degradation: potential limitations of solid lipid nanoparticles. Food Res Int. 2013;52(1):342–9.

Ribeiro HS, Chu B-S, Ichikawa S, Nakajima M. Preparation of nanodispersions containing β-carotene by solvent displacement method. Food Hydrocoll. 2008;22(1):12–7.

Saberi AH, Fang Y, McClements DJ. Stabilization of vitamin E-enriched mini-emulsions: influence of organic and aqueous phase compositions. Colloids Surf A Physicochem Eng Asp. 2014;449(1):65–73.

Sakurai H, Katoh A, Yoshikawa Y. Chemistry and biochemistry of insulin-mimetic vanadium and zinc complexes: trial for treatment of diabetes mellitus. Bull Chem Soc Jpn. 2006;79(11):1645–64.

Sekhon BS. Food nanotechnology—an overview. Nanotechnol Sci Appl. 2010;3:1–15.

Shah B, Davidson PM, Zhong Q. Encapsulation of eugenol using Maillard-type conjugates to form transparent and heat stable nanoscale dispersions. LWT-Food Sci Technol. 2012;49(1):139–48.

Shah BR, Zhang C, Li Y, Li B. Bioaccessibility and antioxidant activity of curcumin after encapsulated by nano and Pickering emulsion based on chitosan-tripolyphosphate nanoparticles. Food Res Int. 2016;89.(Pt 1:399–407.

Shariffa YN, Tan TB, Abas F, Mirhosseini H, Nehdi IA, Tan CP. Producing a lycopene nanodispersion: the effects of emulsifiers. Food Bioprod Process. 2016;98:210–6.

Shariffa YN, Tan TB, Uthumporn U, Abas F, Mirhosseini H, Nehdi IA, Wang YH, Tan CP.Producing a lycopene nanodispersion: formulation development and the effects of high pressure homogenization. Food Res Int. 2017;101:165–72.

Silva HD, Cerqueira MA, Souza BWS, Ribeiro C, Avides MC, Quintas MAC, Coimbra JSR, Carneiro-Da-Cunha MG, Vicente AA. Nanoemulsions of β-carotene using a high-energy emulsification-evaporation technique. J Food Eng. 2011;102(2):130–5.

Son EJ, Lee JS, Lee M, Vu CHT, Lee H, Won K, Park CB. Self-adhesive graphene oxide-wrapped TiO2 nanoparticles for UV-activated colorimetric oxygen detection. Sensors Actuators B Chem. 2015;213:322–8.

Sotelo-Boyás M, Correa-Pacheco Z, Bautista-Baños S, Gómez y Gómez Y. Release study and inhibitory activity of thyme essential oil-loaded chitosan nanoparticles and nanocapsules against foodborne bacteria. Int J Biol Macromol. 2017;103:409–14.

Sozer N, Kokini JL. Nanotechnology and its applications in the food sector. Trends Biotechnol. 2009;27(2):82–9.

Sun J, Wang F, Sui Y, She Z, Zhai W, Wang C, Deng Y. Effect of particle size on solubility, dissolution rate, and oral bioavailability: evaluation using coenzyme Q(10) as naked nanocrystals. Int J Nanomedicine. 2012;7:5733–44.

Surh J, Decker EA, McClements DJ. Utilisation of spontaneous emulsification to fabricate lutein-loaded nanoemulsion-based delivery systems: factors influencing particle size and colour. Int J Food Sci Technol. 2017;52(6):1408–16.

Tan CP, Nakajima M. β-Carotene nanodispersions: preparation, characterization and stability evaluation. Food Chem. 2005;92(4):661–71.

Tan TB, Chu WC, Yussof NS, Abas F, Mirhosseini H, Cheah YK, Nehdi IA, Tan CP.Physicochemical, morphological and cellular uptake properties of lutein nanodispersions prepared by using surfactants with different stabilizing mechanisms. Food Funct. 2016a;7(4):2043–51.

Tan TB, Yussof NS, Abas F, Mirhosseini H, Nehdi IA, Tan CP. Stability evaluation of lutein nanodispersions prepared via solvent displacement method: the effect of emulsifiers with different stabilizing mechanisms. Food Chem. 2016b;205:155–62.

Tang D-W, Yu S-H, Ho Y-C, Huang B-Q, Tsai G-J, Hsieh H-Y, Sung H-W, Mi F-L. Characterization of tea catechins-loaded nanoparticles prepared from chitosan and an edible polypeptide. Food Hydrocoll. 2013;30(1):33–41.

Teng Z. Food protein-based nanoparticles as bioavailability enhancing encapsulants. University of Maryland, MD, USA; 2015.

Teo A, Lee SJ, Goh KKT, Wolber FM. Kinetic stability and cellular uptake of lutein in WPI-stabilised nanoemulsions and emulsions prepared by emulsification and solvent evaporation method. Food Chem. 2017;221:1269–76.

Tsagkaris AS, Tzegkas SG, Danezis GP. Nanomaterials in food packaging: state of the art and analysis. J Food Sci Technol. 2018;55(8):2862–70.

Vasile C, Râpă M, Ştefan M, Stan M, Macavei S, Darie-Niţă RN, Barbu-Tudoran L, Vodnar DC, Popa EE, Ştefan R, Borodi G, Brebu M. New PLA/ZnO:Cu/Ag bionanocomposites for food packaging. Express Polym Lett. 2017;11(7):531–44.

Vejdan A, Ojagh SM, Adeli A, Abdollahi M. Effect of TiO_2 nanoparticles on the physico-mechanical and ultraviolet light barrier properties of fish gelatin/agar bilayer film. LWT-Food Sci Technol. 2016;71:88–95.

Walker RM, Gumus CE, Decker EA, McClements DJ. Improvements in the formation and stability of fish oil-in-water nanoemulsions using carrier oils: MCT, thyme oil, & lemon oil. J Food Eng. 2017;211:60–8.

Wang P, Liu H-J, Mei X-Y, Nakajima M, Yin L-J. Preliminary study into the factors modulating β-carotene micelle formation in dispersions using an in vitro digestion model. Food Hydrocoll. 2012;26(2):427–33.

Wang W, Sun C, Mao L, Ma P, Liu F, Yang J, Gao Y. The biological activities, chemical stability, metabolism and delivery systems of quercetin: a review. Trends Food Sci Technol. 2016;56:21–38.

Wang T, Bae M, Lee JY, Luo Y. Solid lipid-polymer hybrid nanoparticles prepared with natural biomaterials: a new platform for oral delivery of lipophilic bioactives. Food Hydrocoll. 2018;84:581–92.

Wissing SA, Müller RH. A novel sunscreen system based on tocopherol acetate incorporated into solid lipid nanoparticles. Int J Cosmet Sci. 2001;23(4):233–43.

Xu J, Yang F, Chen L, Hu Y, Hu Q. Effect of selenium on increasing the antioxidant activity of tea leaves harvested during the early spring tea producing season. J Agric Food Chem. 2003;51(4):1081–4.

Xu F, Pandya JK, Chung C, McClements DJ, Kinchla AJ. Emulsions as delivery systems for gamma and delta tocotrienols: formation, properties and simulated gastrointestinal fate. Food Res Int. 2018;105:570–9.

Yin L-J, Chu B-S, Kobayashi I, Nakajima M. Performance of selected emulsifiers and their combinations in the preparation of β-carotene nanodispersions. Food Hydrocoll. 2009;23(6):1617–22.

Youssef AM, El-Sayed SM. Bionanocomposites materials for food packaging applications: concepts and future outlook. Carbohydr Polym. 2018;193:19–27.

Zambrano-Zaragoza ML, Mercado-Silva E, Del Real L A, Gutiérrez-Cortez E, Cornejo-Villegas MA, Quintanar-Guerrero D. The effect of nano-coatings with α-tocopherol and xanthan gum on shelf-life and browning index of fresh-cut "Red Delicious" apples. Innov Food Sci Emerg Technol. 2014;22:188–96.

Zhou H, Liu G, Zhang J, Sun N, Duan M, Yan Z, Xia Q. Novel lipid-free nanoformulation for improving oral bioavailability of coenzyme Q10. Biomed Res Int. 2014;2014:793–879.

Chapter 5
Nanobiotechnology in Food Packaging

5.1 Introduction

Customers today demand a lot more from packaging in terms of protecting the quality, freshness and safety of foods. The nanotechnology, which uses microscopic particles, is effective and affordable and will bring out suitable food and dairy packaging in the near future. Food packaging is considered to be one of the earliest commercial applications of nanotechnology in the food sector. Applications for food contact materials using nanotechnology is: incorporating nanomaterials to improve packaging properties (flexibility and gas barrier properties); active packaging that incorporates nanoparticles with antimicrobial or oxygen scavenging properties; intelligent or smart food packaging incorporating nanosensors for sensing and signaling of microbial and biochemical changes; release of antimicrobials, antioxidants, enzymes, flavours and nutraceuticals to extend shelf life; and biodegradable polymer–nanomaterial composites by introduction of inorganic particles, such as clay, into the biopolymeric matrix and can also be controlled with surfactants that are used for the modification of layered silicate.

Food packaging has a key effect on maintaining quality and safety of foods, especially fruits and vegetables during storage as well as transportation and delivery to customers. In fact, packaging has the potential to prolong the shelf life of the food by limiting undesirable factors and conditions such as chemical contaminants, moisture, microorganisms and enzyme activities. Type and concentration of packaging materials have main effects on efficacy of the provided packaging films to extend the shelf life of the packed food products. Currently, petroleum-based plastic materials, such as polyethylene terephthalate and polypropylene, are widely used in food packaging, which have crucial environmental problems and concerns. Therefore, biodegradable and natural polymers as an alternative to petroleum-based packaging have gained much attention in the past decade. Figure 5.1 shows numerous biodegradable polymers which are used in food packaging (Assa et al. 2015).

H. Jafarizadeh-Malmiri et al., *Nanobiotechnology in Food: Concepts, Applications and Perspectives*, https://doi.org/10.1007/978-3-030-05846-3_5

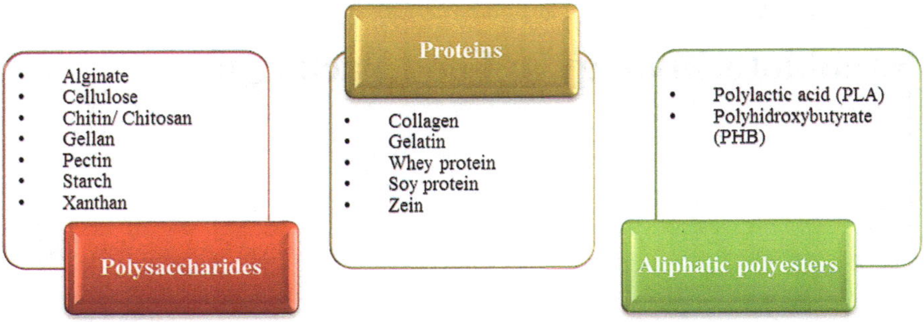

Fig. 5.1 Different biopolymers used in food packaging (Assa et al. 2015)

The combination of food packaging materials and active substances is a new way to control surface microbial contamination of foods. These active food contact materials can extend the product shelf life and enhance food quality and safety that ultimately results in less food waste. Various food grade antimicrobial agents are suggested to utilize in edible food packaging such as sorbic acid, potassium sorbate, benzoic acid, citric acid, polysaccharides, such as chitosan, and nanoparticles, especially metal and metal oxide nanoparticles. Among these, chitosan offers real potential for application in these kinds of packaging systems (Dutta et al. 2009; Joerger 2007; Leceta et al. 2013; Rhim et al. 2013). The classification and summary of this food packaging conjugated with nanoparticles are studied in this chapter.

5.2 Active Packaging

Biopolymers alone cannot be used specially for food packaging, which needs mechanical and barrier properties. Therefore, mechanical and thermal properties of biopolymers can be improved by the addition of additives as filler.

Active packaging is a novel packaging method which is incorporated with materials such as antioxidant, antimicrobial, pigment, nutrients and nanoparticles to make an interaction between packaging and food to improve food quality and safety. This method can employ some technology to delay oxidation and control the respiration rate, moisture migration and microbial growth to prolong the shelf life of food products. Table 5.1 shows several active packaging systems with their applications in food.

Among the mentioned active packaging systems, antioxidant and antimicrobial packaging systems are widely utilized in food packaging. Antioxidant packages prevent oxidation of the food product by two methods: firstly, releasing of antioxidant compound to food, and secondly scavenging unfavorable substances like oxygen. Application of antioxidant packaging, which releases antioxidant materials to food, has benefits related to the direct addition of antioxidant to food that in this

Table 5.1 Some active packaging systems used in food packaging (Assa et al. 2015)

Type of application	Foods
Oxygen scavengers	Ground coffee, tea, roasted nuts, potato chips, chocolate, fat powdered milk, powdered drinks, bread, tortillas, pizza, pizza crust, refrigerated fresh pasta, fruit tortes, cakes, cookies, beer, deli meats, smoke and cured meats, fish, cheese
Carbon dioxide absorbers	Ground coffee
Carbon dioxide emitters	Meat, fish
Moisture absorbers	Dry and dehydrated products, meat, poultry, fish
Ethylene scavengers	Kiwifruit, banana, avocados, persimmons
Ethanol emitters	Bread, cakes, fish
Antimicrobial releasing films	Dry apricots
Antioxidant releasing films	Cereals
Flavor absorbing films	Navel orange juice
Flavor releasing films	Ground coffee
Color containing films	Surimi
Anti-fogging films	Some fresh fruit and vegetable packages
Anti-sticking films	Soft candies, cheese slices
Light absorbers	Pizza, milk
Time-temperature indicators	Microwaveable pancake syrup, refrigerated pasta, deli items

method these active materials release to food by controlled concentration rates and its determined concentrations before storage maintain constant (Peng et al. 2013; Restuccia et al. 2010; Woranuch and Yoksan 2013).

Another kind of active packaging which has recently attracted attention is the antimicrobial active packaging. The main function of this packaging is to assist the preservation and to prolong the shelf life of packaged food by suppressing the growth of spoilage microorganisms (Appendini and Hotchkiss 2002). The active agents for production of antimicrobial packaging include organic acids, fungicides, bacteriocins, enzymes, planet extracts and organic and inorganic nanoparticles. The usage of these agents relates to the kind of microorganisms which causes the spoilage of packaged food. Organic acids, bacteriocins and spice extracts show good influence on control of meat spoilage (Chen and Brody 2013; Quintavalla and Vicini 2002). Potassium sorbate and nisin have antimicrobial activity against yeast and gram-positive bacteria, respectively (Lee et al. 2003; Pranoto et al. 2005). Application of layered silicate fillers, such as montmorillonite, will improve the mechanical and thermal as well as gas barrier properties of bio polymers (Cushen et al. 2012).

Silver nanoparticles (AgNPs) ensure safety of food and can preserve food for longer periods by killing the microorganisms when they are used in the packaging system. Additionally, packaging films and coatings which have AgNPs are able to adsorb and decompose ethylene, which is a natural plant hormone produced during ripening process (Mohammadlou et al. 2016).

Therefore, active packaging does not only provide a barrier against harmful external factors, but also releases or absorbs substances into food and its environment to maintain or even improve the condition of food products.

5.3 Intelligent Packaging

Intelligent food contact materials are mainly intended to monitor the condition of packaged food or the environment surrounding the food. With a visible indicator, this technology can inform the supplier or consumer that foodstuffs are still fresh, kept at the appropriate temperatures throughout the supply chain or the packaging has been breached. Intelligent packaging systems can be divided into three categories, namely time temperature indicators, gas indicators and thermochromic inks (Silvestre et al. 2011). Time temperature indicators, which began appearing on some food products in the late twentieth century, allow suppliers to confirm that the foods have been kept at the appropriate temperature. They fall into two categories: one relies on the migration of a dye through a porous material, which is temperature and time dependent; the other makes use of a chemical reaction (initiated when the label is applied to the packaging), which results in a color change. These indicators allow consumers to feel confident about what they are purchasing and manufacturers to trace their foods along the supply line (Taoukis and Labuza 1989).

Food is a complicated material to package because it is capable of respiration and therefore may change its own atmosphere when inside a package. The gas composition within a package can easily change due to the interaction of food with its environment. Gas indicators are helpful means of monitoring the composition of gases inside a package by producing a change in the color of the indicator through a chemical or enzymatic reaction. Several gas indicators have been developed to detect oxygen, water vapor, ethanol and hydrogen sulfide (De Jong et al. 2005).

Inks are available that are temperature sensitive and can change color based on temperature. These inks can be printed onto packages or labels such that a message can be conveyed to the consumer based on the color of the ink they are seeing. Thermochromic inks can let a consumer know whether a foodstuff is hot or cold. Thermochromic inks are becoming a popular technology for beverages. The inks used can be adversely affected by UV light and temperature over 121 °C, and therefore consumers should not fully rely on the ink message when it comes to deciding the proper time to consume the food (Robertson 2006).

Micropores and sealing defects in packaging systems can lead food products to an unexpected high exposure to oxygen, which can result in undesirable changes. Nanoparticles can be applied as reactive particles in packaging materials to inform about the state of the package. The so-called nanosensors are able to respond to environmental changes (e.g. temperature or humidity in storage rooms, levels of oxygen exposure), degradation products or microbial contamination (Ahvenainen 2003).

The recent developments for smart packaging include oxygen indicators, freshness indicators and pathogen sensors. Oxygen allows aerobic microorganisms to grow during food storage. There has been an increasing interest in developing non-toxic and irreversible oxygen sensors to assure oxygen absence in oxygen-free food packaging systems such as packaging under vacuum or nitrogen (Silvestre et al. 2011).

5.4 Biodegradable Coatings and Films

Biodegradable coatings are thin layers of edible component applied to the fruit's surface in addition to or as a replacement for natural protective waxy coatings that act as a physical barrier towards carbon dioxide, oxygen and moisture movement for the fruits. Coatings can be formulated from different components such as hydrocolloids (polysaccharides and proteins), lipids (waxes and resins) and synthetic polymers. These edible materials have different barrier properties against gases and physico-chemical and mechanical characteristics. Therefore, most coatings are made of more than one material with the addition of low molecular weight molecules including sorbitol, polyols or glycerol that serve as plasticizers. Biodegradable coatings are applied on fruits and vegetables to improve appearance, delay ripening, reduce water loss and decay, and extend shelf life, but may also change flavor. In fact, semi-permeable coatings can modify the internal atmosphere of fruits by changing the composition or concentration of standard atmosphere gases, like controlled atmosphere storage, with less expense incurred. The atmosphere created by coatings is strongly related to the permeability of the coating and fruit respiration rate and can alter as function of environmental conditions including temperature and humidity (Jafarizadeh Malmiri et al. 2011a, b, 2012a, b). Biodegradable films are the prepared films by natural biopolymers which are wrapped around the fruits and vegetables and create a modified atmosphere within them and delay their ripening process (Berenjian et al. 2015).

The polysaccharides as coating materials for fruits have been used extensively in the past few years. Cellulose and its derivatives, such as methyl cellulose, carboxymethyl cellulose, hydroxypropyl cellulose and Hydroxypropyl methylcellulose, and chitosan as a cationic and antimicrobial polysaccharide, have been widely used in the formulation of edible coatings and films (Berenjian et al. 2015; Jafarizadeh Malmiri et al. 2011a, b).

Incorporation of some antimicrobial agents into the coating and film formulations can lead to an effective antimicrobial activity, promoting the fruit's microbial stability throughout storage. Benzoates, sorbates, short-chain organic acids (e.g. acetic, lactic, propionic, fumaric and citric acids) and fungicides (e.g. benomyl, thiabendazole, iprodione and captan) have been widely used to prepare antimicrobial coatings and films (Jafarizadeh Malmiri et al. 2012a, b). The typical structure of an antimicrobial edible coating is shown in Fig. 5.2.

Moreover, nanotechnology provides food scientists with a number of ways to create novel laminate films suitable for use in the food and dairy industry. One of the most powerful methods is based on the layer by layer deposition technique, in which the charged surfaces are coated with interfacial films consisting of multiple nanolayers of different materials (Jafarizadeh Malmiri et al. 2012a, b).

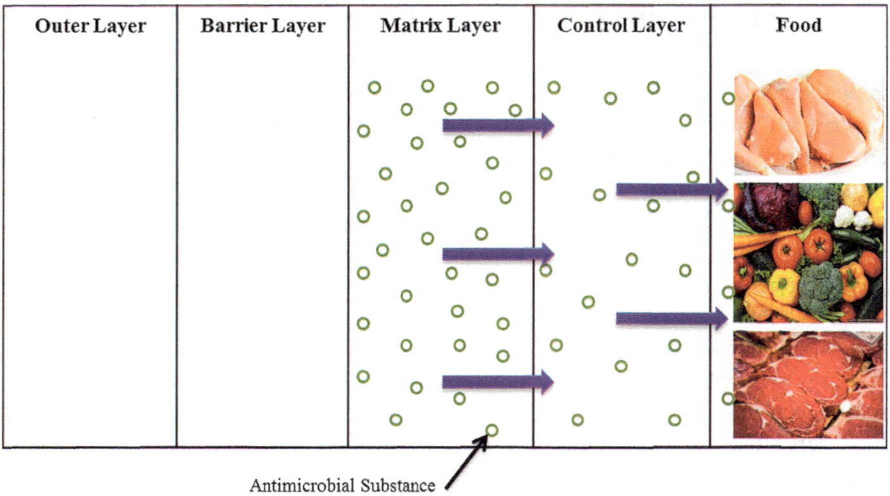

Fig. 5.2 Structure of an antimicrobial coating (Assa et al. 2015)

5.5 Nanoparticles in Food Packaging

Although the application of nanotechnology in food and beverages areas are mostly currently at R&D or near-market stages, its applications in food packaging has already become a commercial reality. Food packaging applications form the largest share of the current and short-term predicted market for nano-enabled products in the food sector. When integrated into food packaging, nanosensors can detect specific indicators of pathogen metabolism or can alert or inform the consumer about a product's temperature, light or oxygen exposure history. This could eliminate the need for expiry dates in some incidences and may even give the consumer a more accurate estimation of the state of spoilage of the food. The combination of food packaging materials and active substances is a new way to control surface microbial contamination of foods. Some nanomaterials exhibit antimicrobial effects. For such active packaging material, sharing a common interface or physical contact with the food surface is essential. These active food contact materials can extend the product shelf life, enhancing food quality and safety and ultimately leading to less food waste (Méndez-Vilas 2013).

5.5.1 Organic Nanoparticles

Nowadays, there is an increasing interest in natural polymers which have inherently various properties for active packaging. Due to their biodegradable characteristics, they can be nominated as an alternative to commodity polymers. In addition, since they have intrinsic antimicrobial and antioxidant properties, there is no need for incorporating active agents. Most of these polymers are made from proteins and polysaccharides which can be replaced by some conventional synthetic packaging materials exerted to preserve food products. Organic materials, which naturally exist in foods such as carbohydrates, fat, vitamins and proteins, can be in different sizes such as large macromolecules or a simple mono molecule in the nano range size. Due to special properties of biological nanomaterials, they can be used for various purposes to improve taste, texture and sustainability. Efficiency of these nanomaterials depends on different parameters such as their biological and chemical properties and composition of the food (Suppakul et al. 2003).

Polylactic acid, a common biodegradable nanoparticle, is often used to encapsulate and deliver micronutrients like iron, vitamin and protein. It has shown that the polylactic acid needs an associative compound such as polyethylene glycol for successful results, and the functional ingredients can be encapsulated in nanoparticles and released in response to specific environmental triggers (Sundarraj et al. 2014).

Among active packaging sources, chitosan is a biopolymer with special properties. Chitosan as a polysaccharide, which is derived by deacetylation of chitin, is obtained from crab and shrimp shells. Chitosan is a biopolymer which consists of β-(1-4)-2-acetamido-D-glucose and β-(1-4)-2-amino-D-glucose units as shown in Fig. 5.3. It can be applied in scientific and industrial fields like biotechnology, pharmaceutics, cosmetics, wastewater treatment, drug delivery, food science and packaging. Its significant characteristics, such as biofunctionality, biodegradability, nontoxic, film forming and antimicrobial activity, have made chitosan a promising packaging material. Besides these characteristics, chitosan films are able to postpone the rate of mass transfer, moisture and shelf life (Assa et al. 2017; Jafarizadeh Malmiri et al. 2012a, b).

Potential applications of chitin and chitosan in numerous areas is shown in Fig. 5.4. Chitosan nanoparticles present more effective properties than bulk chitosan due to their higher surface area to volume ratio. As mentioned, the antimicrobial property of chitosan is one of its major characteristics which make it suitable for application in food packaging. By applying nanoparticles instead of a large amount of chitosan in food packaging, the charge density will increase which provides a higher interaction by the surface of microbe's cell. Furthermore, the size reduction of chitosan to its nanoparticles improves their dispersion in biopolymers and improves their efficiency as an antimicrobial agent in food packaging. Chitosan nanoparticles can be fabricated by various methods such as emulsion droplet coalescence, emulsion cross linking, ionotropic gelation, template polymerization and molecular self-assembly, which results in changes in shape and particle size (Assa et al. 2017; Jafarizadeh Malmiri et al. 2012a, b).

Ionic gelation, also known as ion-induced gelation, is a mild method for chitosan NP preparation in an aqueous media without any chemical surface modification of the chitosan molecule. Furthermore, this method requires simple operating conditions (Kunjachan et al. 2014; Vaezifar et al. 2013).

In the chitosan NPs preparation using ionic gelation method, the counter-ions have been widely utilized. These compounds can be categories in three main groups,

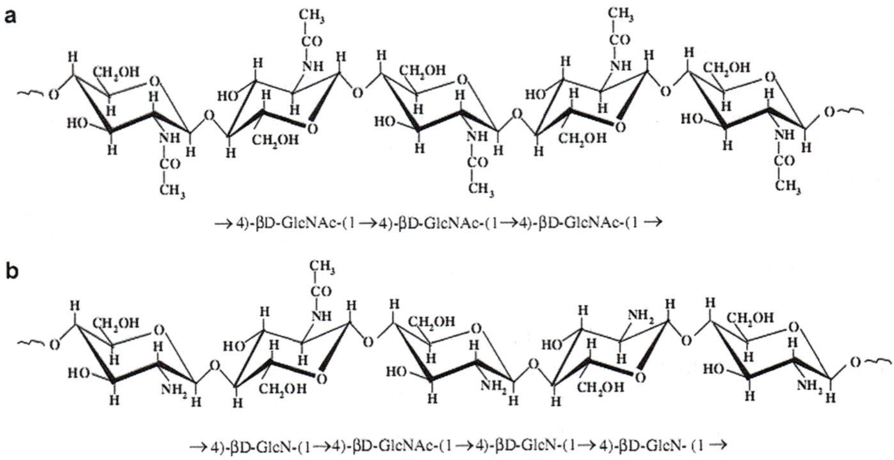

Fig. 5.3 Chemical structure of (**a**) chitin and (**b**) chitosan (Jafarizadeh Malmiri et al. 2012a, b)

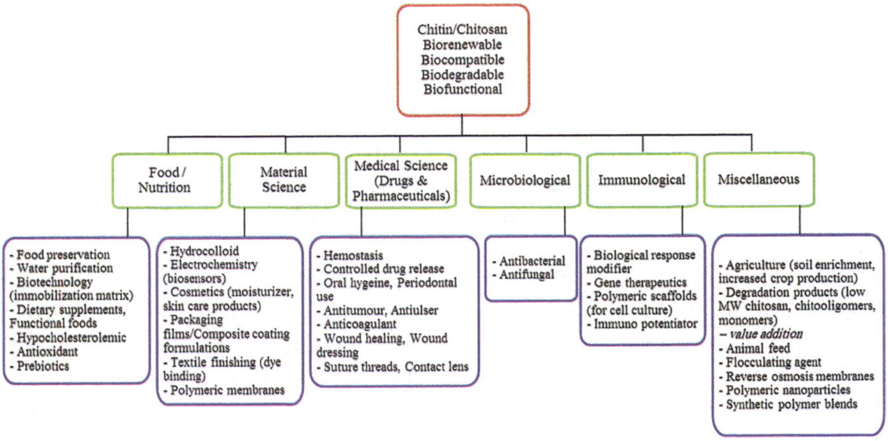

Fig. 5.4 Potential application of chitin and chitosan (Assa et al. 2015)

namely low molecular weight materials (e.g. pyrophosphate and tripolyphosphate), hydrophobic counter-ions (e.g. alginate and polyaldehydro-carbonic acid) and high molecular weight ions (e.g. octyl sulphate and cetostearyl sulphate) (Sinha et al. 2004). Tripolyphosphate (TPP) is a non-toxic polyanion which its negatively charged ions can be interacted with chitosan protonated amine groups through electrostatic forces and create ionic cross-linked networks. The particle size of the chitosan NPs using ionic gelation method and TPP as counter-ions is strongly affected by several parameters such as concentration of chitosan and TPP, and reaction time (Du et al. 2009; Martins et al. 2012).

5.5.2 Inorganic Nanoparticles

Metal nanomaterials received substantial attention in this area, and products utilizing the unique antimicrobial properties of metals, such as silver and gold, have been launched onto the market. Metal oxides have also been incorporated into commercialized products displaying light activated microbe inactivation. Ag nanoparticles absorb and decompose ethylene, thus food packaging films which incorporate Ag impart this effect on the associated food. This may contribute to its positive effects on the shelf life of fruits and vegetables by inhibiting senescence. Side effects are reported to be rare when AgNPs are applied at doses permitted by the FDA (Méndez-Vilas 2013). The antimicrobial efficacy of AgNPs has been studied against a number of bacteria, among them those of food safety concern, too. It was also found that AgNPs are effective against silver resistant strains such as *E. coli*. Inorganic nanoparticles, like AgNPs, can easily be incorporated into polymers while keeping their antimicrobial activity (Mohammadlou et al. 2016). In Europe, carbon black and silica dioxide nanomaterials have already been used as food contact materials (Higashisaka et al. 2015).

The antimicrobial action of chitosan can be improved by the addition of some NPs to chitosan films. The addition of AgNPs to chitosan were probed against *S. aureus,* and it was concluded that their minimum inhibitory concentrations were 500-fold lower than bulk chitosan (Ali et al. 2011). ZnO nanofibers, which are inherently antimicrobial, were mixed with chitosan and the resulted composite was used against *E. coli* and *Candida albicans*. It was observed that *E. coli* and *C. albicans* growth were completely prevented at low minimum inhibitory concentrations (Wang et al. 2012).

A UV-activated colorimetric oxygen indicator, which uses nanoparticles of TiO_2 to photosensitize the reduction of methylene blue by triethanolamine in a polymer encapsulation medium using UVA light, had been developed and used (Lee et al. 2002). Upon UV irradiation, the sensor bleaches and remains colorless, until it is exposed by oxygen, when its original blue color is restored. The rate of color recovery is proportional to the level of oxygen exposure. Nanocrystalline SnO_2 as a photosensitizer in a colorimetric O_2 indicator has been used with the color of the film varying depending on the O_2 exposure (Mills and Hazafy 2009). Also pH indicators based on organically modified silicate nanoparticles have been recently introduced (Jurmanović et al. 2010; Yam 2000).

References

Ahvenainen R. Novel food packaging techniques. Elsevier; 2003.

Ali SW, Rajendran S, Joshi M. Synthesis and characterization of chitosan and silver loaded chitosan nanoparticles for bioactive polyester. Carbohydr Polym. 2011;83(2):438–46.

Appendini P, Hotchkiss JH. Review of antimicrobial food packaging. Innovative Food Sci Emerg Technol. 2002;3(2):113–26.

Assa F, Jafarizadeh-Malmiri H, Anarjan N, Berenjian A, Ghasemi Y. Applications of chitosan nanoparticles in active biodegradable and sustainable food packaging. In: Kale SA, Durai PRT, Prabakar K, editors. Renewable energy and sustainable development. Hauppauge: Nova Science Publishers; 2015.

Assa F, Jafarizadeh-Malmiri H, Ajamein H, Vaghari H, Anarjan N, Ahmadi O, Berenjian A. Chitosan magnetic nanoparticles for drug delivery systems. Crit Rev Biotechnol. 2017;37(4):492–509.

Berenjian A, Ebrahiminezhad A, Southee R, Jafarizadeh-Malmiri H. Design and investigation on antimicrobial properties for novel chitosan based active food packaging. Rep Opinion. 2015;7(7):1–3.

Chen J, Brody AL. Use of active packaging structures to control the microbial quality of a ready-to-eat meat product. Food Control. 2013;30(1):306–10.

Cushen M, Kerry J, Morris M, Cruz-Romero M, Cummins E. Nanotechnologies in the food industry—recent developments, risks and regulation. Trends Food Sci Technol. 2012;24(1):30–46.

De Jong A, Boumans H, Slaghek T, Van Veen J, Rijk R, Van Zandvoort M. Active and intelligent packaging for food: is it the future? Food Addit Contam. 2005;22(10):975–9.

Du W-L, Niu S-S, Xu Y-L, Xu Z-R, Fan C-L. Antibacterial activity of chitosan tripolyphosphate nanoparticles loaded with various metal ions. Carbohydr Polym. 2009;75(3):385–9.

Dutta P, Tripathi S, Mehrotra G, Dutta J. Perspectives for chitosan based antimicrobial films in food applications. Food Chem. 2009;114(4):1173–82.

Higashisaka K, Yoshioka Y, Tsutsumi Y. Applications and safety of nanomaterials used in the food industry. Food Safety. 2015;3(2):39–47.

Jafarizadeh Malmiri H, Osman A, Tan C, Abdul R. Development of an edible coating based on chitosan-glycerol to delay 'Berangan' banana (Musa sapientum cv. Berangan) ripening process. Int Food Res J. 2011a;18(3).

Jafarizadeh Malmiri H, Osman A, Tan C, Rahman AR. Evaluation of effectiveness of three cellulose derivative-based edible coatings on changes of physico-chemical characteristics of 'Berangan' banana (Musa sapientum cv. Berangan) during storage at ambient conditions. Int Food Res J. 2011b;18(4):1381.

Jafarizadeh Malmiri H, Osman A, Tan CP, Abdul Rahman R. Effects of edible surface coatings (sodium carboxymethyl cellulose, sodium caseinate and glycerol) on storage quality of Berangan banana (Musa sapientum cv. Berangan) using response surface methodology. J Food Process Preserv. 2012a;36(3):252–61.

Jafarizadeh Malmiri H, Jahanian MAG, Berenjian A. Potential applications of chitosan nanoparticles as novel support in enzyme immobilization. Am J Biochem Biotechnol. 2012b;8(4):203–19.

Joerger RD. Antimicrobial films for food applications: a quantitative analysis of their effectiveness. Packag Technol Sci. 2007;20(4):231–73.

Jurmanović S, Kordić Š, Steinberg MD, Steinberg IM. Organically modified silicate thin films doped with colourimetric pH indicators methyl red and bromocresol green as pH responsive sol–gel hybrid materials. Thin Solid Films. 2010;518(8):2234–40.

Kunjachan S, Jose S, Lammers T. Understanding the mechanism of ionic gelation for synthesis of chitosan nanoparticles using qualitative techniques. Asian J Pharm (AJP). 2014;4(2).

Leceta I, Guerrero P, Cabezudo S, de la Caba K. Environmental assessment of chitosan-based films. J Clean Prod. 2013;41:312–8.

Lee S-W, Mao C, Flynn CE, Belcher AM. Ordering of quantum dots using genetically engineered viruses. Science. 2002;296(5569):892–5.

Lee CH, An DS, Park HJ, Lee DS. Wide-spectrum antimicrobial packaging materials incorporating nisin and chitosan in the coating. Packag Technol Sci. 2003;16(3):99–106.

Martins AF, de Oliveira DM, Pereira AG, Rubira AF, Muniz EC. Chitosan/TPP microparticles obtained by microemulsion method applied in controlled release of heparin. Int J Biol Macromol. 2012;51(5):1127–33.

Méndez-Vilas A. Microbial pathogens and strategies for combating them: science, technology and education. Badajoz: Formatex Research Center; 2013.

Mills A, Hazafy D. Nanocrystalline SnO2-based, UVB-activated, colourimetric oxygen indicator. Sens Actuators B Chem. 2009;136(2):344–9.

Mohammadlou M, Maghsoudi H, Jafarizadeh-Malmiri H. A review on green silver nanoparticles based on plants: synthesis, potential applications and eco-friendly approach. Int Food Res J. 2016;23(2):446–63.

Peng Y, Wu Y, Li Y. Development of tea extracts and chitosan composite films for active packaging materials. Int J Biol Macromol. 2013;59:282–9.

Pranoto Y, Rakshit S, Salokhe V. Enhancing antimicrobial activity of chitosan films by incorporating garlic oil, potassium sorbate and nisin. LWT Food Sci Technol. 2005;38(8):859–65.

Quintavalla S, Vicini L. Antimicrobial food packaging in meat industry. Meat Sci. 2002;62(3):373–80.

Restuccia D, Spizzirri UG, Parisi OI, Cirillo G, Curcio M, Iemma F, Puoci F, Vinci G, Picci N. New EU regulation aspects and global market of active and intelligent packaging for food industry applications. Food Control. 2010;21(11):1425–35.

Rhim J-W, Park H-M, Ha C-S. Bio-nanocomposites for food packaging applications. Prog Polym Sci. 2013;38(10–11):1629–52.

Robertson G. Modified atmosphere packaging. Food packaging–principles and practice. 2nd ed. Florida: CRC Press; 2006.

Silvestre C, Duraccio D, Cimmino S. Food packaging based on polymer nanomaterials. Prog Polym Sci. 2011;36(12):1766–82.

Sinha V, Singla AK, Wadhawan S, Kaushik R, Kumria R, Bansal K, Dhawan S. Chitosan microspheres as a potential carrier for drugs. Int J Pharm. 2004;274(1–2):1–33.

Sundarraj A, GuhanNath S, Aaron S, Ranganathan T. Recent innovations in nanotechnology in food processing and its various applications—a review. Int J Pharm Sci Rev Res. 2014;29(2):116–24.

Suppakul P, Miltz J, Sonneveld K, Bigger SW. Active packaging technologies with an emphasis on antimicrobial packaging and its applications. J Food Sci. 2003;68(2):408–20.

Taoukis P, Labuza T. Applicability of time-temperature indicators as shelf life monitors of food products. J Food Sci. 1989;54(4):783–8.

Vaezifar S, Razavi S, Golozar MA, Karbasi S, Morshed M, Kamali M. Effects of some parameters on particle size distribution of chitosan nanoparticles prepared by ionic gelation method. J Clust Sci. 2013;24(3):891–903.

Wang Y, Zhang Q, Zhang C-L, Li P. Characterisation and cooperative antimicrobial properties of chitosan/nano-ZnO composite nanofibrous membrane. Food Chem. 2012;132(1):419–27.

Woranuch S, Yoksan R. Eugenol-loaded chitosan nanoparticles: II. Application in bio-based plastics for active packaging. Carbohydr Polym. 2013;96(2):586–92.

Yam KL. Intelligent packaging for the future smart kitchen. Packag Technol Sci. 2000;13(2):83–5.

Chapter 6
Nano-sensors in Food Nanobiotechnology

6.1 Introduction

Food safety is a very important issue in the food industry and agriculture because it is directly related to the influence of food on human health. Recent food safety and public health concerns about food additives and chemical residues in the food industry have attracted considerable attention. Conventional detection methods for dangerous materials and bacterial pathogens are time consuming and expensive, so it needs to improve rapid, sensitive methods to detect these materials. In recent years, nanotechnology has emerged as a promising field for solving food safety issues in terms of detecting toxin materials. There is good potential for the application of nanosensors for monitoring food quality and safety in the food and agriculture industry. The classification and summary of nanosensor and biosensor applications are studied in this chapter.

Food safety is so important for human health. With an increased variety of foods, this suggests that food quality evaluation and safety in all steps of agricultural food products needs to be considered. All steps include 'from the farm to the kitchen' production, processing, distribution, and marketing of agricultural food to the final consumers (Omanović-Mikličanina and Maksimović 2016). The term food quality includes physical and chemical properties of food, for example the appearance, taste, smell, nutritional value, freshness, flavor, and texture (Mehrotra 2016). Developing control systems ensures us of agricultural food products' quality and safety.

In 1962, biosensors were born with the progress of enzyme electrodes by the scientist Leland C. Clark. A biosensor is a powerful and innovative analytical instrument involving physical, chemical and biological sensing techniques (Lin et al. 1997). The 'bio' and 'sensor' elements can be coupled together in different ways: Matrix Entrapment, Membrane Entrapment, Physical Adsorption, and Covalent Bonding (Fig. 6.1) (Lin et al. 1997; Mehrotra 2016).

H. Jafarizadeh-Malmiri et al., *Nanobiotechnology in Food: Concepts, Applications and Perspectives*, https://doi.org/10.1007/978-3-030-05846-3_6

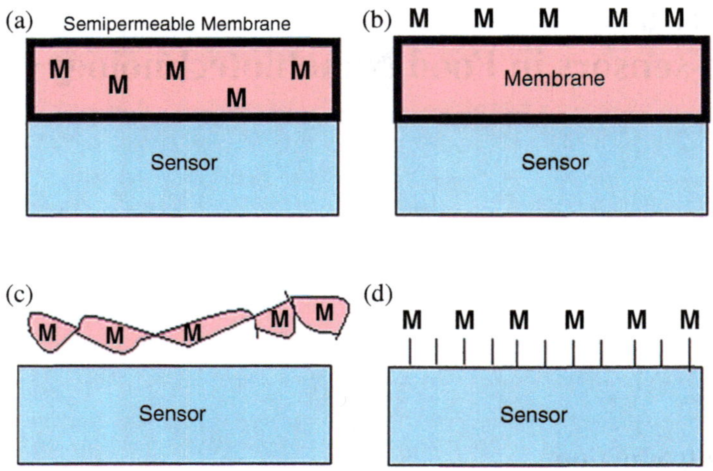

Fig. 6.1 Coupling of Bio-Material (M) with sensor. (**a**) Membrane entrapment. (**b**) Physical absorption. (**c**) Matrix entrapment. (**d**) Covalent bonding

Biosensors have been developed to identify a diversity of biomolecular complexes, including oligonucleotides, antibody-antigen, antibody, antigen, living cells, tissue interactions, hormone-receptor interactions, enzyme-substrate interactions and lectin-glycoprotein interactions. Their applications are in medicine, food, agriculture and biotechnology (Agrawal and Rathore 2014; Mohanty 2001). Nano biosensors were developed for their applications in detection of molecules like urea, glucose, various microorganisms and pathogens. Nanosensors had the advantage over other sensors due to being small, transportable, sensitive, measurable, reliable, exact, reproducible and stable (Agrawal and Rathore 2014).

In general, biosensors have two important ingredients. One is a highly specific recognition element and the other is a transducer that changes the biological response into a calculable signal (Lin et al. 1997).

6.2 Historical Developments of Bio Nanosensors

Several requirements are necessary for developing a biosensor system to assure commercial success (1–6):

1. Selectivity: The biosensor devices should have high selectivity for the goal analyte molecules and no crossing for other molecules with similar chemical structure.
2. Sensitivity: The biosensor devices should be capable to measure the proportion of positives that are correctly identified with minimum steps such as pre concentration of the models.

3. Linearity of response: The linear response range of the system should be measured.
4. Reproducibility of signal response: Same concentrations of samples must have the same response when they are evaluated several times.
5. Quick response time and recovery time: The biosensor device response must be quick, therefore, real time monitoring of the goal analyte molecule and recovery time should be low for reusability of the biosensor devices.
6. Stability and operating life: Different biochemical and environmental conditions cause instability of most of the biological compounds. The biological component used should be stable for a long time to create a useful commercial and practical device (Thakur and Ragavan 2013).

Developments of bio nanosensors are described in detail:

1. First generation biosensors

   Leland Charles Clark Jr. was the pioneer of the biosensor study. In 1956 he showed biosensor components in his first paper: the electrode which measured oxygen concentration in blood. In 1962 the model was illustrated by experiments that glucose oxidase was entrapped at a Clark oxygen electrode using dialysis membrane. The work of Clark was modified by Updike and Hicks in 1967. They established the first functional enzyme electrode based on glucose oxidase immobilized onto an oxygen sensor. In 1970, the first potentiometric enzyme electrode was introduced by Guilbault and Montalvo. It was a urea-sensor based on enzyme urease immobilized at an ammonium selective liquid membrane electrode. Guilbault and Lubrano defined glucose and a lactate enzyme sensor based on hydrogen peroxide detection at a platinum electrode in 1973. A heat-sensitive enzyme sensor was modified by Klaus Mosbach in 1974. In 1975, Lubbers and Opitz extended the concept to make an optical biosensor for alcohol.
2. Second generation biosensors

   In 1976 an electrochemical glucose biosensor in a bedside artificial pancreas was incorporated by Clemens et al. A novel semi-continuous catheter-based blood glucose analyzer was introduced by VIA Medical. Later in 1976 the Lactate Analyzer LA 640 was introduced by La Roche that was used to transport electrons from lactate dehydrogenase to an electrode.
3. Third generation biosensors

   These types of biosensors have advanced into the employment of enzymes and mediators onto the same electrode instead of freely diffusing mediators in the electrolyte. Direct interaction between the enzymes and the electrode was possible so either mediator or enzyme was not required. In 1983, dependency reactions in real time using surface Plasmon resonance (SPR) technique was observed by Liedberg. In 1987 blood-glucose was measured with a pen-sized by Cambridge, USA. The electronics were modified and MediSense's sales showed high growth reaching US$175 million by 1996. The global biosensor market reached $12 billion by the year 2015 and showed enormous potential for advancement and development (Thakur and Ragavan 2013).

6.3 Classification of Biosensors

A biosensor contains three main components: bioreceptor molecules, a biotransducer component, and an electronic system which corporates amplifier, processor or recorders. The bioreceptor can categorize the target analyte molecules because it is a biological molecular species, while the transducer converts a chemical or biochemical signal into an electrical signal. So this integration causes to measure the target analyte molecules without using reagents as shown in Fig. 6.2 (Thakur and Ragavan 2013).

The classification of biosensor is based on their bio-receptor or transducer which includes the following types (see Fig. 6.3). We will discuss the various types of possible biosensors briefly (Mohanty 2001) which are shown in Fig. 6.3.

6.3.1 Electrochemical Biosensors

Electrochemical biosensors are employed for the revelation of hybridized DNA, DNA-binding drugs and glucose concentration. Chemical reactions produce or consume ions or electrons that are considered the fundamental of these biosensors which, in turn, cause some changes in the electrical properties of the solution (Kovacs 1998; Sethi and Lowe 1990).

6.3.2 Potentiometric Biosensor

In these types of sensors, the measured parameter is oxidation/reduction potential (of an electrochemical reaction). The working principle of that is when a ramp voltage is applied to an electrode in solution and the current flow occurs because of electrochemical reaction. The voltage at which these reactions occur indicate a particular reaction and particular species (Sethi and Lowe 1990).

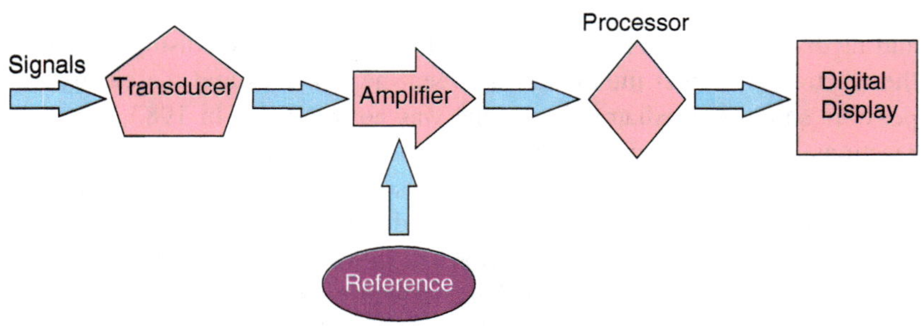

Fig. 6.2 Schematic representation of biosensor components

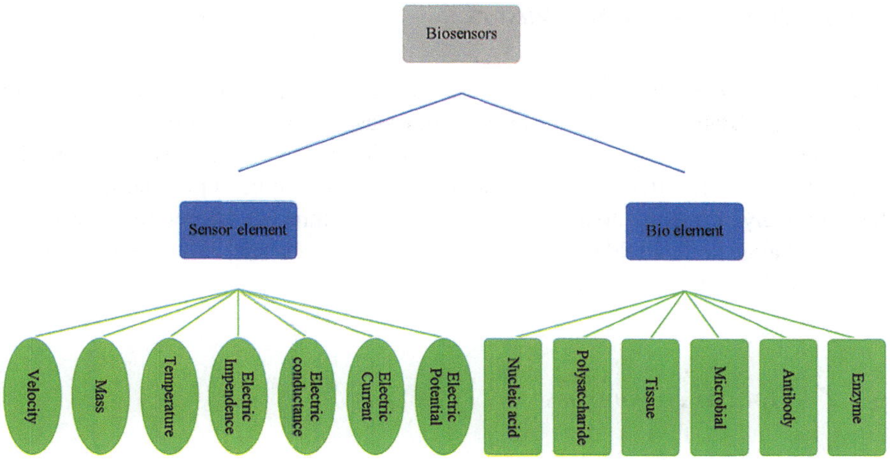

Fig. 6.3 Elements of a biosensor

6.3.3 Calorimetric Biosensors

These kinds of biosensors are produced by incorporating enzymes with temperature sensors. The heat reaction of the collision of enzyme and analyte molecules is measured when the analyte molecules contact with the enzyme and heat is calibrated against the analyte concentration (Kovacs 1998).

6.3.4 Amperometric Biosensor

These biosensors are highly sensitive and are used to detecting electroactive species presented in biological test samples. Enzymes need to catalyze the production of radio-active species because of not electro-active inherently of the biological test samples. In these sensors, current is considered as the measured parameter (Sethi and Lowe 1990).

6.3.5 Resonant Biosensors

These kind of biosensors are produced by incorporating an acoustic wave transducer with antibody. By changing the membrane mass due to attaching the analyte molecules to the membrane, as a result, the resonant frequency of the transducer will be changed that this resonant frequency change is measured out (Kovacs 1998).

6.3.6 *Ion-Sensitive Biosensors*

These biosensors have ion-sensitive surface (Semiconductor Field Effect Transistors). The surface electrical potential variates by interaction between ions and a semiconductor. By covering the sensor electrode using a polymer layer, the Ion Sensitive Field Effect Transistor can be created. This type of polymer layer must be selectively permeable to analyte ions that diffuse through the polymer layer and, in turn, change in the Field Effect Transistor's surface potential. This kind of biosensor is mainly used for pH detection (Fraser 1994).

6.3.7 *Optical Biosensors*

These types of biosensors have light signals as the output transduced signal. The fundamental of creation of these biosensors is an optical diffraction or electrochemilluminence. In optical biosensors, a protein coats a silicon wafer through covalent bonds. The wafer is exposed to UV light through a photomask and the antibodies made inactivated in the exposed regions. The roasted wafer chips are made in active regions when incubated in analyte antigen-antibody binding, so generating diffraction grating. By irradiating a light source, such as laser, this grating produces a diffraction signal. This signal can be measured or can be improved earlier to improve sensitivity (Kovacs 1998).

6.4 Advantages and Types of Nanosensors in the Food Industry

In 2004 an article was published which examined the prospects of nanosensors based on nanotechnology. Unlike the previous article, it emphasized available products and those close to market (Bogue 2008). The nanosensors consist of an inert, biofriendly matrix in which a sensing component and/or an optical reporter are entrapped. Therefore, in this section, we focus on development nanosensors that are currently in use. In particular, we discuss advantages and types of nanosensors. The numerous advantages are listed:

1. Nanosensors can operate cooperative interactions between analytes, enzymes and reporter dyes. They are not limited to using an analyte, and therefore the number of measured analytes can be increased.
2. Food was protected from any pollutant and toxic effects of the sensing dye and enzymes by the matrix of nanosensors.
3. Sensing dye was protected from potential interferents in the cellular environment by the matrix.

4. There is no selective sequestration of nanosensors into food.
5. Nanosensors improve ability to do ratiometric measurements.
6. The in-vitro calibration of nanosensors can be used for in-vivo measurements. It should be considered that they are different from labeled nanoparticles that were modified by a reporter molecule on the outside of them (Aylott 2003).

In each section we first describe the type of nanosensors and explain the properties related to the nanosensor. We do not provide a complete overview of the available nanosensors, but we describe the current type of the nanosensors and their applications in the food industry based on the type of detection (Riu et al. 2006). Mostly, two kinds of nanosensors, for example chemical nanosensors and bio nanosensors, can be used in the food industry. Chemical sensors and biosensors are produced in a similar way by putting chemical or bacterial detectors (reagents) on the probe, which in reaction of material with the targeted chemical element or bacteria (Ileš et al. 2011).

Sometimes optical nanosensors are used in the food industry with chemical or bio detection. Their color changes when they react with chemical or biological reagents (Ileš et al. 2011). Figure 6.4 shows types of used nanosensors in the food industry that are described below.

6.4.1 Nanosensors Based on Nanostructure

Nanoparticles are particles and atoms at the nanoscale. Because of their size, which is of the same order as the de Broglie wavelength related with the valence electrons, nanoparticles perform zero-dimensional quantum dots with separate energy levels. A quantum dot is a location that can contain a single electrical charge (electron). The properties of a quantum dot were changed with presence or absence of an electron, and they can be applied for several purposes such as to save data or transducers in nanosensors. Nanoparticles have prominent size-dependent optical properties that optical nanosensors are built based on noble metal nanoparticles or semiconductor

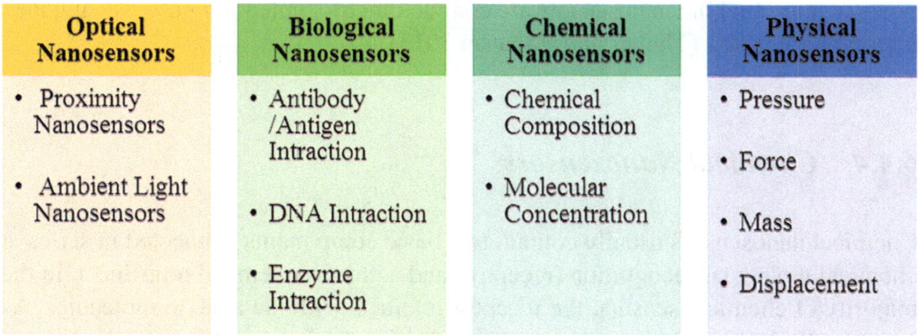

Fig. 6.4 Types of nanosensors

quantum dots. When nanostructures of a smaller size than the de Broglie wavelength for electrons create an intense absorption in the UV-visible region as absent in the spectrum of the bulk material. When analyte molecules are attached to the nanoparticles, the local refractive index changes. To detect molecules attached to the noble metal nanoparticles, a shift in the UV-visible spectrum can be used. The selectivity of the sensor is obtained by a chemical development of nanoparticles with self-assembled monolayers that can be incorporated in a wide variety of molecular components such as enzymes, antibodies or DNA (Riu et al. 2006).

6.4.2 Optical Nanosensors

The optical nanosensors include a spectrofluorometer or a fluorescence microscope with quantitative fluorescence detection that can be used for their analytical performance. The nanosensors are suspended in solution and known amounts of the analyte were added. Nanosensors are suspended in an aqueous solution so the aqueous phase does not scatter light and seems transparent because of the small size of the nanosensors. Sasaki et al. introduced the first optical nanosensor based on fluorescein entrapped in a polyacrylamide nanoparticle and applied to pH sensing (Aylott 2003).

6.4.3 Bio Nanosensors

Bio nanosensors are an analytical device for the measurement of a specific analyte. They contain biological material and physicochemical transducers to enable rapid, accurate and sensitive detection of goal, for example electrochemical, optical and thermometric. Conventional microbial biosensors include the metabolic utilities of the microorganisms to detect a matter that is either a substrate or an inhibitor of the processes. The construction of enzyme sensors are complicated and expensive, and therefore microbial biosensors are better than enzyme biosensors. Some of the main types of microbial bio nanosensors are conductometric, potentiometric and amperometric nanosensors (Thakur and Ragavan 2013).

6.4.4 Chemical Nanosensors

Chemical nanosensors usually contain two basic components connected in series: a chemical molecular recognition (receptor) and a physicochemical transducer. In the majority of chemical sensors, the receptor interacts with the analyte molecules. As a result, the physical properties are altered in such a way that the appending transducer can gain an electrical signal.

6.4.5 Physical Nanosensors

The physical nanosensors sense the variation environmental physical such as force, acceleration, flow rate, mass, volume, density and pressure.

6.4.6 Sol-Gel Nanosensors

Sometimes mineral oxide sol-gel has been used as the matrix for optical nanosensors. Sol-gel matrix has unique properties such as being porous, highly pure, optically transparent, chemically inert, photo stable and thermally stable, making it an ideal matrix for quantitative spectrophotometric measurements. Moreover, the flexible processing situations permit for modifying of sol-gel properties, such as pore size and hydrophilic-hydrophobic balance, depending on what sensing molecules are to be immobilized (Aylott 2003).

6.5 Nanosensors in the Agri-Food Industry

Other study indicated that the potential for synthesized nanoparticles to irreversibly attach to targeted bacteria thus inhibiting them from binding and infecting their host and decreasing the infectious contamination. Still, there is a need for a significant amount of research in this area to create the opportunities for improvement and benefits (Magnuson et al. 2011).

The potential for nanotechnology was studied to revolutionize agricultural and food systems in several areas by a National Planning Workshop in the U.S. Nanosensors were one of the opportunities discussed for improving the security of the food supply. In 2003 concepts in nano science that were essential for understanding food properties and developing interventions to change the structural and functional attributes of food were studied by Leser, Michel, and Watzke (Sanguansri and Augustin 2006).

Nanotechnology in the food industry was compared with numerous other industries. This technology has shown great potential in this industry. According to Taylor Research, the nanotechnology industry grew to US$2.6 trillion in manufactured products by the year 2014. Moreover, according to the U.S. Department of Agriculture, by 2015, the global impact of nano products will be US$1 trillion annually (Durán and Marcato 2013). Recently, nanoparticles can also be applied as reactive particles in packaging materials. These nanoparticles, or nanosensors, are designed to respond to environmental condition changes, for example temperature, pH or moisture, during storage due to degradation products of the food by microorganisms (Bouwmeester et al. 2009).

In 2008 Fu et al. modified biosensors that attached fluorescent dye particles to anti-salmonella antibodies on nanorod array. Dye particles on the sensor become visible if there are bacteria in the food. This biosensor can detect the bacteria and toxins in food and crops unlike the time-consuming conventional lab tests (bacterial cultures). The main objective of the design of bio nanosensors is to decrease detection time from days to hours. Nanoparticles or nanosensors also conserve food products and act as anti-bacterial action in packaging food and prevent growth of pathogens during storage (Bouwmeester et al. 2009).

Nanosensors, in conjunction with polymers, are used to monitor food pathogens and chemicals during storage and transit processes in smart packaging (Pathakoti et al. 2017).

6.6 Mechanism of Biosensors

In food microbiology, nanosensors or nanobiosensors are used for the detection of pathogens in processing plants or in food material, quantification of available food constituents and alerting consumers and distributors on the safety status of food. Nanosensors have been modified for food analysis, flavors, drinking water and clinical application (Pathakoti et al. 2017; Singh et al. 2017). A low-cost nano bioluminescent spray was used in food for detection of microbes, which microbes react with it and obtained a visual glow, thereby indicating microbial contamination using it. Pathogens can be detected by nano cantilevers, and recently bacterial growth of *Escherichia coli* are detected by a developed rapid biosensor (Pathakoti et al. 2017).

However, it is believed that nanoparticles for active packaging, such as nano-silver, nano-gold, titanium-oxide and zinc-oxide, release into the food and react with microbes (García et al. 2010; Singh et al. 2017; Thakur and Ragavan 2013). Some of the significant bio nanosensors and their mechanism of action are listed in Table 6.1.

Table 6.1 Bio nanosensors and their mechanism and analytes

Type of bio nanosensor	Mechanism	Analyte	Reference
Polarographic oxygen electrode	Electrode based	Oxygen in Blood	Clark (1956)
Electrochemical	Enzyme electrode	Glucose	Updike and Hicks (1967)
Amperometric	Electrode with immobilized glucose oxidase	Blood glucose	Guilbault and Lubrano (1973)
Amperometric	Dual enzyme electrode system	Organophosphorus pesticides	Gouda et al. (2001)
Aptasensors	Aptamer	Vitamin B_{12}	Selvakumar and Thakur (2012)

6.7 Nanosensor Applications

Nanosensors are used in many clinical or non-clinical applications. Available nanosensor types and their potential applications are summarized in Fig. 6.5. One of the most important applications of nanosensors is in the food and agriculture industries. Using nanoparticles or nanosensors in food and agriculture products, such as food packaging, food additives and food preservation, is one of the important applications of nanotechnology in the food industry. Improvement of food safety and extending product life is the basic goal of application in the food and agriculture industry, which includes nanomaterials with antibacterial properties and nanosensors for microorganisms and toxin material detection for increasing product life (Ileš et al. 2011). The main aim of nanosensors is to reduce the time for pollutant detection to minute (Sozer and Kokini 2009). Figure 6.6 and Table 6.2 show the potential nanosensors have in the food and agriculture sector.

6.8 Disadvantage of Bio Nanosensors

Nanosensors have great benefits in the agriculture and food industry; however, there is an enormous public concern about potential toxicity and their environmental effects. Studies on animals have shown potential toxicity and the uncertainty of the behavior of nanoparticles in the human body. Therefore, standardized studies need to be established to evaluate the effects of nanoparticles on human health and

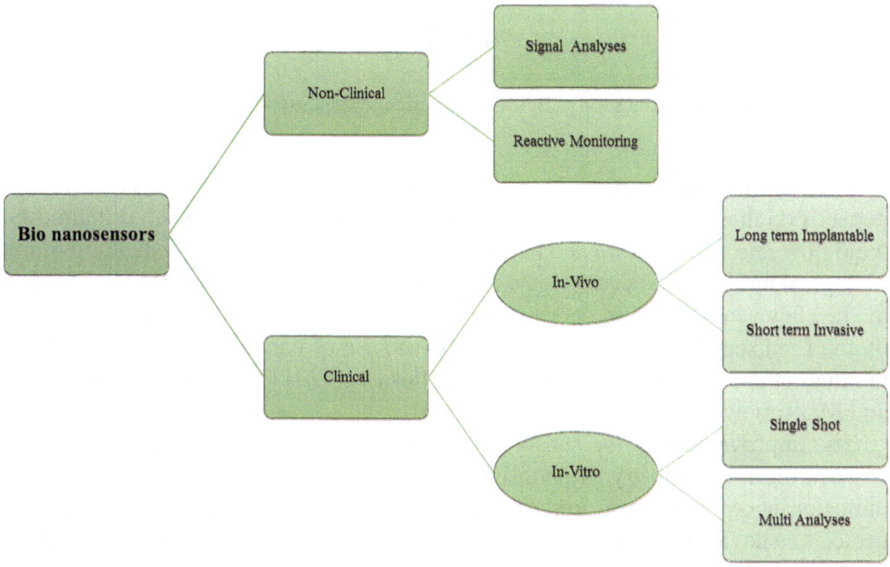

Fig. 6.5 Potential application of Bio nanosensors

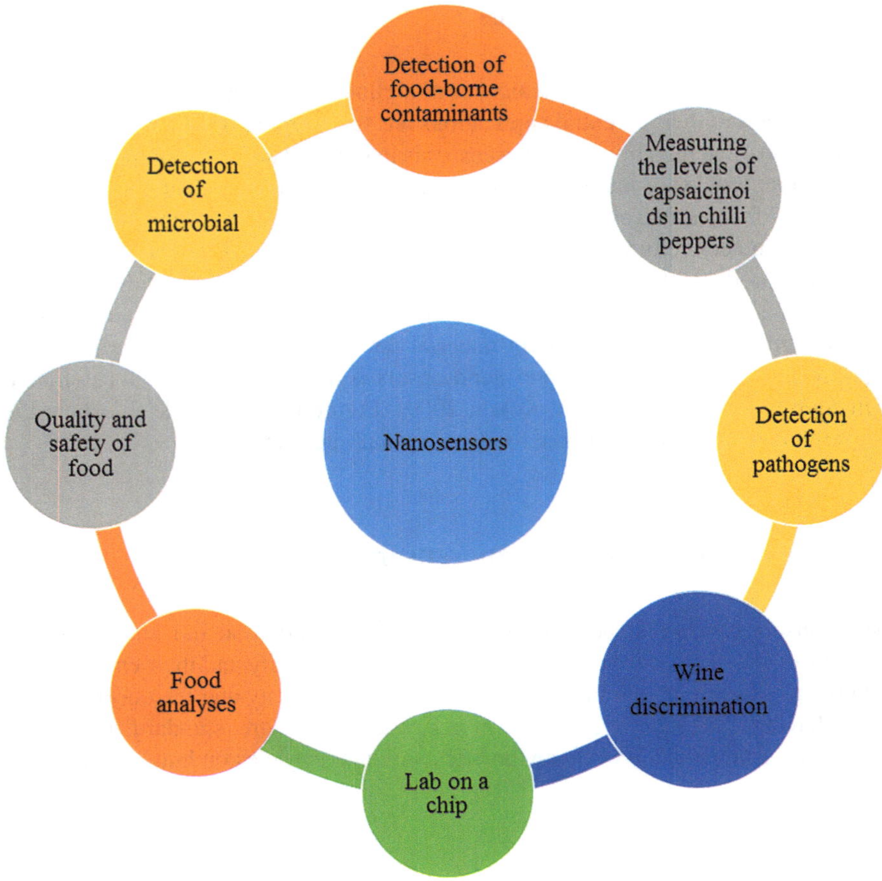

Fig. 6.6 Application of nanosensors in food and agriculture industry

their environmental impact. The toxicity of nanoparticles in the environment depends on their size, type, charge and environmental influences such as humidity, wind flow rate, temperature and nature of light. Therefore, the understanding of the biological and toxicological effects of nanomaterials has significantly advanced in the last few years. It is also necessary to establish a set of protocols and regulations on the food security of IP implications. So the use of nanotechnology depends on the market and the geographical position (Ileš et al. 2011; Omanović-Mikličanina and Maksimović 2016).

The improvement of the sensors, nanosensors and indicators are an important issue in the food industry. A nanosensor measures only certain features, while an indicator incorporates measurement and monitoring. The nanosensors must be connected to a device for signal transduction of the receptor, while an indicator directly provides qualitative or semi-quantitative information about the quality for a visible change (Fuertes et al. 2016).

Table 6.2 Nanosensors' potential application in the agriculture and food industry

Agriculture	Food Processing	Food Packing	Food Transport	Nutrition
• Nanosensors for monitoring soil conditions (e.g. moisture, soil pH) • Nanosensors for detection of foodborne contaminants or for monitoring environmental conditions at the farm • Nanochips for identity preservation and tracking • Nanocapsules for delivery of pesticides, herbicides, fertilizers and vaccines • Nanosensors and nanobased smart delivery systems for efficient use of agricultural natural resources • Nanoparticles to deliver growth hormones or DNA to plants in controlled manner • Nanoparticles used as smart nanosensors for early warning of changing conditions	• Nanoencapsulated flavor enhancers • Nanotubes and nanoparticles as gelation and viscosifying agents • Nanocapsule infusion of plant based steroids to replace a meat's cholesterol • Nanoparticles to selectively bind and remove chemicals or pathogens from food • Aptasensors for determination of microbial toxins	• Portable nanosensors to detect chemicals, pathogens and toxins in food • Nanosensors incorporated into packaging materials for detection of chemicals released during food spoilage and serve as electronic tongue • Nanosensors to detect ethylene • Nanosensors applied as labels or coating to add an intelligent function to food packaging to microbial safety • Aptasensors for determination of microbial cells, antibiotics, drugs and their residues and heavy metals	• Nanosensors for monitoring environmental conditions during storage • Nanosensors for traceability and monitoring product conditions during transport and storage • Smart-sensor technology for monitoring the quality of grain, dairy products, fruit and vegetables in a storage environment • Aptasensors for determination of microbial cells	• Nanocapsules incorporated into food to deliver nutrients • Nanocochleates for delivering nutrients without affecting the color or taste of food

References

Agrawal S, Rathore P. Nanotechnology pros and cons to agriculture: a review. Int J Curr Microbiol App Sci. 2014;3(3):43–55.

Aylott JW. Optical nanosensors—an enabling technology for intracellular measurements. Analyst. 2003;128(4):309–12.

Bogue R. Nanosensors: a review of recent progress. Sensor Rev. 2008;28(1):12–7.

Bouwmeester H, Dekkers S, Noordam MY, Hagens WI, Bulder AS, De Heer C, Ten Voorde SE, Wijnhoven SW, Marvin HJ, Sips AJ. Review of health safety aspects of nanotechnologies in food production. Regul Toxicol Pharmacol. 2009;53(1):52–62.

Clark L. Monitor and control of blood and tissue oxygen tensions. Trans Am Soc Artif Intern Organs. 1956;2:41–8.

Durán N, Marcato PD. Nanobiotechnology perspectives. Role of nanotechnology in the food industry: a review. Int J Food Sci Technol. 2013;48(6):1127–34.

Fraser D. Glucose biosensors: the sweet smell of success. Med Device Technol. 1994;5:44.

Fuertes G, Soto I, Carrasco R, Vargas M, Sabattin J, Lagos C. Intelligent packaging systems: sensors and nanosensors to monitor food quality and safety. J Sensors. 2016;2016:8p.

García M, Forbe T, Gonzalez E. Potential applications of nanotechnology in the agro-food sector. Food Sci Technol. 2010;30(3):573–81.

Gouda M, Thakur M, Karanth N. Stability studies on immobilized glucose oxidase usingan amperometric biosensor—effect of protein based stabilizing agents. Electroanalysis. 2001;13(10):849–55.

Guilbault G, Lubrano G. An enzyme electrode for the amperometric determination of glucose. Anal Chim Acta. 1973;64(3):439–55.

Ileš D, Martinović G, Kozak D. Review of potential use, benefits and risks of nanosensors and nanotechnologies in food. Strojarstvo. 2011;53(2):127–36.

Kovacs GT. Micromachined transducers sourcebook. New York: WCB/McGraw-Hill; 1998.

Lin VS-Y, Motesharei K, Dancil K-PS, Sailor MJ, Ghadiri MR. A porous silicon-based optical interferometric biosensor. Science. 1997;278(5339):840–3.

Magnuson BA, Jonaitis TS, Card JW. A brief review of the occurrence, use, and safety of food-related nanomaterials. J Food Sci. 2011;76(6):R126–33.

Mehrotra P. Biosensors and their applications—a review. J Oral Biol Craniofac Res. 2016;6(2):153–9.

Mohanty SP. Biosensors: a survey report. Citeseer; 2001.

Omanović-Mikličanina E, Maksimović M. Nanosensors applications in agriculture and food industry. Bull Chem Technol Bosnia Herzegovina. 2016;47:59–70.

Pathakoti K, Manubolu M, Hwang H-M. Nanostructures: current uses and future applications in food science. J Food Drug Anal. 2017;25(2):245–53.

Riu J, Maroto A, Rius FX. Nanosensors in environmental analysis. Talanta. 2006;69(2):288–301.

Sanguansri P, Augustin MA. Nanoscale materials development—a food industry perspective. Trend Food Sci Technol. 2006;17(10):547–56.

Selvakumar L, Thakur M. Nano RNA aptamer wire for analysis of vitamin B12. Anal Biochem. 2012;427(2):151–7.

Sethi R, Lowe C. Electrochemical microbiosensors. In: IEE Colloquium on Microsensors, IET, London; 1990. p. 9/1–9/5.

Singh T, Shukla S, Kumar P, Wahla V, Bajpai VK, Rather IA. Application of nanotechnology in food science: perception and overview. Front Microbiol. 2017;8:1501.

Sozer N, Kokini JL. Nanotechnology and its applications in the food sector. Trend Biotechnol. 2009;27(2):82–9.

Thakur M, Ragavan K. Biosensors in food processing. J Food Sci Technol. 2013;50(4):625–41.

Updike S, Hicks G. The enzyme electrode. Nature. 1967;214(5092):986.

Chapter 7
Nano-encapsulation for Nutrition Delivery

7.1 Introduction

This chapter is aiming to discuss nanoencapsulation of various water-insoluble materials. Nanoencapsulation is a technique to improve the stability of insoluble bioactive compounds in water. Nanoencapsulation not only enhances the aqueous solubility and stability of the bioactive compounds but also provides controlled release to protect their biological/pharmacological activity in the body. At the same time, bioavailability improves by controlling the release, and thus the probability of repeated use can be reduced. Various techniques are used for the nanoencapsulation of bioactive compounds for improving their release in the target site. The most commonly used techniques for compound encapsulation are nanoprecipitation, nano-emulsification, coacervation, spray drying, electrospinning and electrospray, solvent evaporation and other methods that are reviewed in this chapter.

Recently, nanotechnology has developed as a scientific field of research which consequently increases reactivity with changes in mechanical, electrical and optical properties. These properties cause many unique and wide applications in different fields. Nanotechnology is known as the design, characterization and production by controlling shape, decreasing size at the nanoscale and increasing the surface area (Anandharamakrishnan 2014). Food, pharmaceutical and cosmeceutical nanotechnology is an emerging technology with the potential to produce products and processes at an industrial scale. Currently, the main applications of nanotechnology in food and medicine are:

1. Nanocomposites in packaging material (for controlling diffusion and microbial protection),
2. Nanobiosensors (for detection of contamination and quality deterioration),
3. Nanoencapsulation or nanocarriers (Anandharamakrishnan 2014; Reis et al. 2006a, b).

H. Jafarizadeh-Malmiri et al., *Nanobiotechnology in Food: Concepts, Applications and Perspectives*, https://doi.org/10.1007/978-3-030-05846-3_7

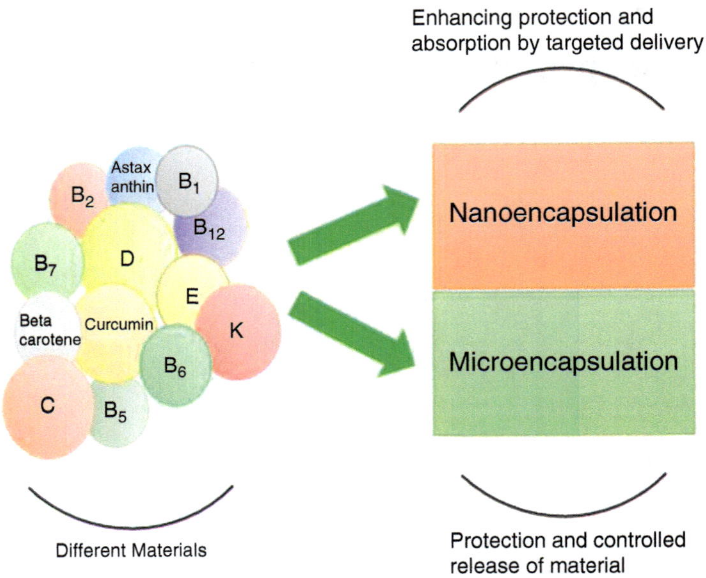

Fig. 7.1 Different materials for nanoencapsulation and microencapsulation

Vitamins, antioxidants and other bioactive compounds are useful for human health as they assist the body to grow. Additionally, they can stop diseases and help general health. Unfortunately, bioactive compounds are sensitive molecules; therefore, they should be preserved from harmful agents like heat, light and oxidants. Encapsulation is a promising and novel method for preserving the innate characteristics of these materials; some of them are shown in Fig. 7.1 (Katouzian and Jafari 2016).

Many reviews and research papers were published on application of nanotechnology in the food and medicine industries. Nevertheless, some research is focused on nanoencapsulation of food and bioactive ingredients (Anandharamakrishnan 2014). Therefore, this chapter gives attention to the discussion of the various nanoencapsulation techniques and their advantages as well as evaluating the interesting emerging technologies and trends in this field, along with the safety and regulatory issues.

7.2 Nanoencapsulation

The unencapsulated or free food or drug ingredients often have low bioavailability which is mainly attributable to the premature degradation and poor solubility in the gastrointestinal tract (Pandey et al. 2005). Encapsulation is a rapidly developing technology with many potential applications in areas such as the pharmaceutical and food industries. Microcapsules are particles which have a diameter from 1 to 5000 μm. Nanoencapsulation of drugs and food ingredients includes forming loaded particles with diameters ranging from 1 to 1000 nm. Nanoencapsulation protects

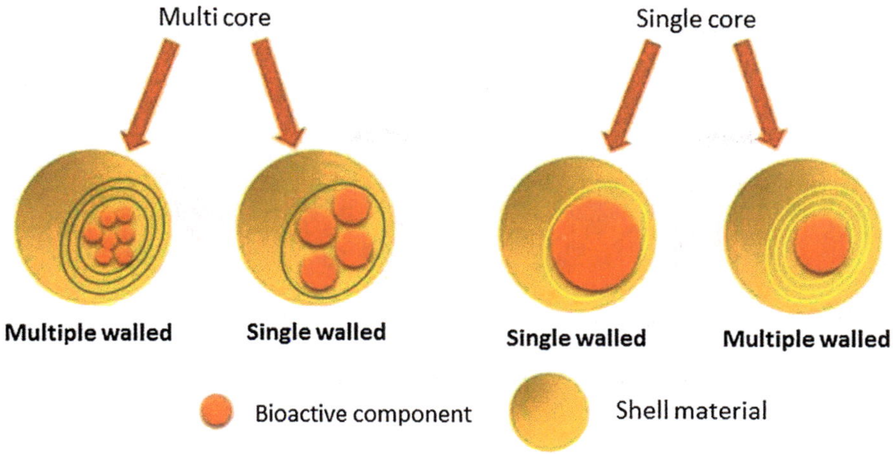

Fig. 7.2 Types of nanostructure delivery systems applied in encapsulation

drug and bioactive compounds, such as polyphenols, micronutrients, enzymes, antioxidants and nutraceuticals, from adverse environmental factors and also for controlled release at the targeted site. Figure 7.2 shows types of nanostructure delivery systems applied in encapsulation (Hădărugă et al. 2008; Reis et al. 2006a).

Nanoparticles for encapsulation have different types like liposomes and solid lipid nanoparticles, polymeric microspheres, micelle, nanogel, nanospheres and nanocapsules. Nanospheres and nanocapsules are shown in Fig. 7.3. Nanocapsules are defined as vesicular systems in which the bioactive compound is confined to a cavity or inner liquid core surrounded by a unique polymer membrane. In nanocapsules, the active substances are usually dissolved in the inner core or may be adsorbed at the surface. Whereas nanospheres have a matrix structure in which the bioactive compound is uniformly dispersed or absorbed in the inner area.

Decreasing the size of the particles in encapsulation to the nanoscale and increasing surface area compared to mass proportion improves bioavailability, enhances controlled release and enables precision targeting of the bioactive compounds to a greater extent than microencapsulation. Moreover, nanoencapsulation systems have many benefits:

- ease of handling,
- improved stability against coalescence and gravitational separations,
- protection against oxidation,
- targeted delivery,
- pass along the smallest body fenestrations,
- retention of volatile ingredients,
- formulating optically transparent bioactive compound solutions,
- taste masking,
- moisture-triggered controlled release,
- pH-triggered controlled release,
- consecutive delivery of multiple active ingredients,

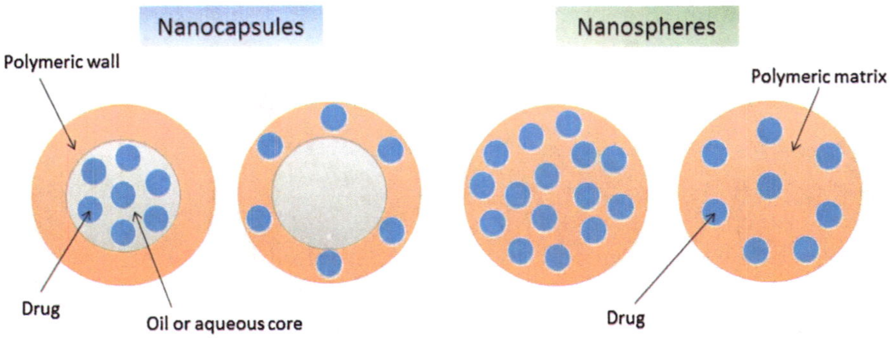

Fig. 7.3 Schematic representation of nanospheres and nanocapsules

- change in flavor character,
- long-lasting organoleptic perception,
- faster dissociation,
- high intracellular uptake,
- enable precision targeting, reduce the reactions between bioactive compounds and other molecules plus surrounding medium,
- reduction in the quantity of utilized core-shell material (Anandharamakrishnan 2014; Katouzian and Jafari 2016; Reis et al. 2006b; Yadav et al. 2011).

Edible nanoencapsulation vehicles, such as various lipid, polysaccharides, proteins and biodegradable polymers, are safe ingredients, biocompatible and biodegradable (Xiao et al. 2017). Nanocapsules were also produced by a complex coacervation method using two solutions of oppositely charged polymers and their electrostatic interactions. Furthermore, these carriers could be developed by adding other polar or magnetism groups to get higher selectivity (Xing et al. 2005).

Micelles are bioactive nanocarriers for the encapsulation of water insoluble agents such as amphiphilic block copolymers, polymers and surfactants. The hydrophobic ends assemble as the core and hydrophilic ends assemble as the outer shell that are able to form spherical micelles in an aqueous solution. Nanogels are aqueous dispersions of hydrogel particles formed by physically or chemically cross-linked polymer networks in the nanoscale size. In a drug delivery system, nanogels have shown their potential as carriers of the active compounds. This is considered to be unprecedented for common pharmaceutical nano-carriers because of their high loading capacity, high stability and responsiveness to environmental factors such as ionic strength, pH and temperature (Nuruzzaman et al. 2016).

7.3 Materials for Nanoencapsulation

Bioactive components, such as antioxidants, decrease oxidation and thus prevent cell damage caused by free radicals so protection of these components is important (O'Connor et al. 2016). The material for encapsulation is commonly related to the

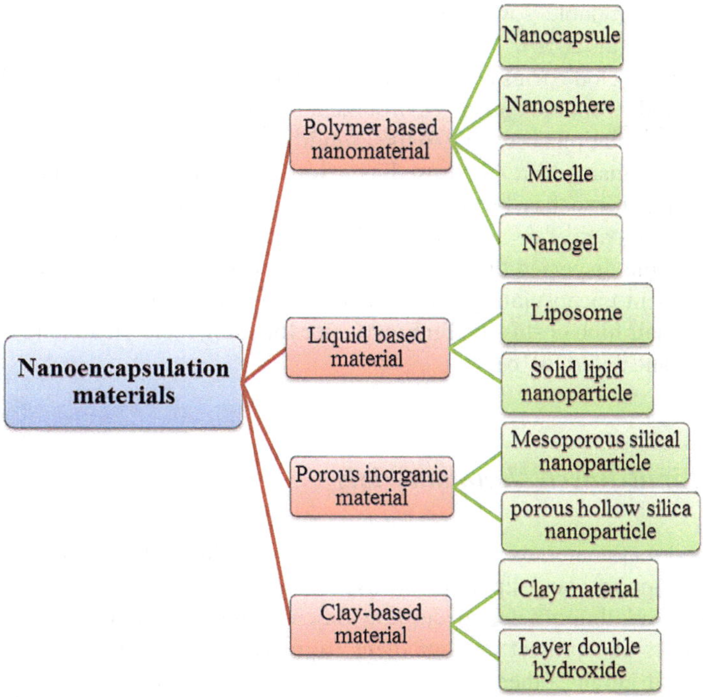

Fig. 7.4 Promising nanoencapsulation materials used

internal phase, the core material, the filler or the fill. The encapsulation material is known as the external phase, shell, coating or membrane, for example nanocapsules. Nanoencapsulation of bioactive compounds involves the formation of bioactive compounds loaded or entrapped particles having a diameter within the nano-range. During encapsulation, a multi-stage delivery pattern can be observed as some bioactive compounds are absorbed and attached to the outer surface of the shell (Nuruzzaman et al. 2016). The protection of bioactive compounds, such as vitamins, antioxidants, proteins and lipids, is obtained by using different nanoencapsulation techniques (Noronha et al. 2013). These materials are known as nanoencapsulation materials and encapsulate the bioactive compounds, for example nanocapsules, nanospheres, micelles, nanogels, liposomes and inorganic nano-cages. Figure 7.4 shows some of materials that are used to encapsulate for delivery purposes.

7.3.1 Polymer-Based Nanoencapsulation Materials

Mostly, polymer nanocomposites, which contain a polymer with nanoparticles or nano-fillers dispersed within the polymer matrix, were used for encapsulating the bioactive compounds. Polymers produced by natural sources are environmentally friendly, biodegradable, low cost and do not produce any degradation byproducts (Nuruzzaman et al. 2016).

Starch and cyclodextrins are carbohydrate-based biopolymers that protect bioactive compounds from the outside environment's conditions. Gum Arabic can be also used for encapsulation according to its solubility, viscosity and emulsification properties. Also alginates have been used as a wall material at environment temperatures. Ethylcellulose has been accepted as a good substance for encapsulating water-soluble bioactive compounds. As the wall materials width rise, the water permeability of the dispersed bioactive compounds decreas. Protein-based shells may also be used in encapsulating different bioactive compounds. Nevertheless, economically, their high cost is the limiting factor for using them on an industrial scale (Katouzian and Jafari 2016).

7.3.2 Lipid-Based Nanomaterials

Lipid-based nanomaterials were considered as potential to encapsulate the hydrophilic, hydrophobic and lipophilic active compounds for delivery systems with higher encapsulating efficiency and low toxicity (Nuruzzaman et al. 2016). Lipid-based carriers can protect an ingredient from free radicals, metal ions, pH and enzymes that might otherwise cause degradation of the bioactive compound (Mozafari et al. 2006).

7.3.3 Porous Inorganic Nanomaterials

The search for an appropriate carrier material or encapsulation with suitable properties for controlled release is a persistent theme in pesticide application. In recent years the synthesis of inorganic porous nanomaterials with interesting hierarchical morphologies has attracted much attention because of their potential application in a wide range of industrial and biological pursuits. (Nuruzzaman et al. 2016).

7.3.4 Clay-Based Nanomaterials

Nanoclays and layered double hydroxides are considered to be economically biodegradable materials in agricultural and environmental applications. These provide excessive opportunities for developing multifunctional nanocarrier materials for applications in living organisms and material sciences (Nuruzzaman et al. 2016). Also, classification of the main nanoencapsulated bioactive compounds is shown in Table 7.1 (Livney 2015).

Table 7.1 Nanoencapsulated bioactive used recently

Bioactive	Example
Vitamin	D, A, C, B_1 B_2 (Folic acid)
Minerals	Calcium
Carotenoid	β-carotene, Astaxanthin Lutein, β-carotene, Lycopene, Canthaxanthin
(Poly)Phenol	Curcumin, Resveratrol Gallic acid, Eugenol Catechin, Doxorubicin Oleoresin
Fatty-acid	Omega-3, Conjugated linolenic acid
Peptides	Whey protein bioactive peptides
Other nutraceuticals	Clove oil, Coenzyme Q10

7.4 Nanoencapsulation Techniques

There are different techniques for the production of nanocapsules for applications in the areas of catalysis, drug delivery, dye dispersion and display media. Techniques are classified top-down and bottom-up (Anandharamakrishnan 2014; Luo and Zhou 2004). In general, these techniques are divided into three main groups:

(a) Physical processes such as spray drying–coating, extrusion, and spray drying;
(b) Physiochemical processes such as simple or complex coacervation and entrapment into liposomes;
(c) Chemical processes: interfacial polymerization and molecular inclusion (Anandharamakrishnan 2014).

In order to select the most suitable encapsulation process for a given purpose, it is necessary to know the required size of the particles, the physicochemical properties of the core material as well as the nature of the substance to encapsulate. Other parameters to be taken into account are the desired or recommended releasing mechanism and associated costs. Techniques for achieving nanoencapsulation are more complex than those used for microencapsulation, mainly due to the fact that it is more difficult to attain a good nanoencapsulation process. This can be justified by the complex morphology of the capsule and core material and the demands of releasing rates. Chemical and physical natures of the core and shell materials, evaluation of their interactions as well as their proportion in the formulation of the capsules are important parameters which determine the final properties of the particles (Quintanilla-Carvajal et al. 2010). Here, we have selected and reviewed some nanoencapsulation techniques. In this section, some of the most important and novel techniques are presented. Nanoencapsulation can be classified based on top-down and bottom-up strategies (Fig. 7.5).

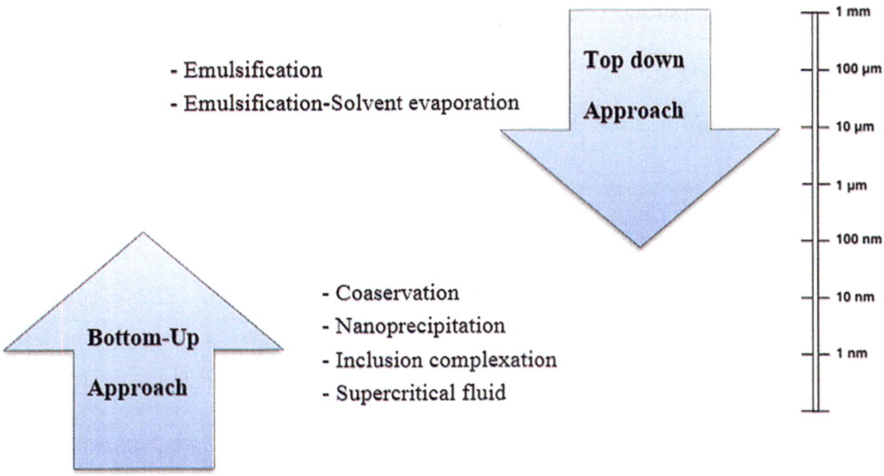

Fig. 7.5 Top-down and bottom-up approaches in nanoencapsulation

7.4.1 Nanoprecipitation Method

The different liquid-based techniques for nanoencapsulation of bioactive compounds are described (Anandharamakrishnan 2014; Nagavarma et al. 2012). The nanoprecipitation method is also known solvent displacement. In this technique, the organic internal phase containing the bioactive compounds is dissolved into the aqueous external phase in an ultrasonic bath. The precipitation of polymer from an organic solution and the diffusion of the organic solvent in the aqueous in the presence or absence of a surfactant occurred in this technique (Fig. 7.6) (Gutiérrez et al. 2013; Reis et al. 2006a).

The effect of several formulation factors can be investigated on the nanocapsules' properties such as mean diameter, zeta potential and entrapment efficiency. Emulsification-solvent evaporation is a modified type of solvent evaporation technique which includes emulsification of the polymer solution into an aqueous. Then, the solvent is evaporated, and the polymer precipitation remains as the nanocapsule. Also, precipitation of nanocapsules was achieved by centrifugation. The size of the particles can be modified by adjusting the stirring rate, type and the amount of dispersing substance, viscosity of the phases and temperature. High-speed homogenization and ultrasonication are applied in order to obtain a small particle size (Amjadi et al. 2013; Katouzian and Jafari 2016).

7.4.2 Nanoemulsification Method

Most common uses of emulsion technology are in aqueous solutions, and nanoemulsions are produced in aqueous medium. The nanoemulsification method is an effective technique to encapsulate both hydrophilic and lipophilic bioactive compounds

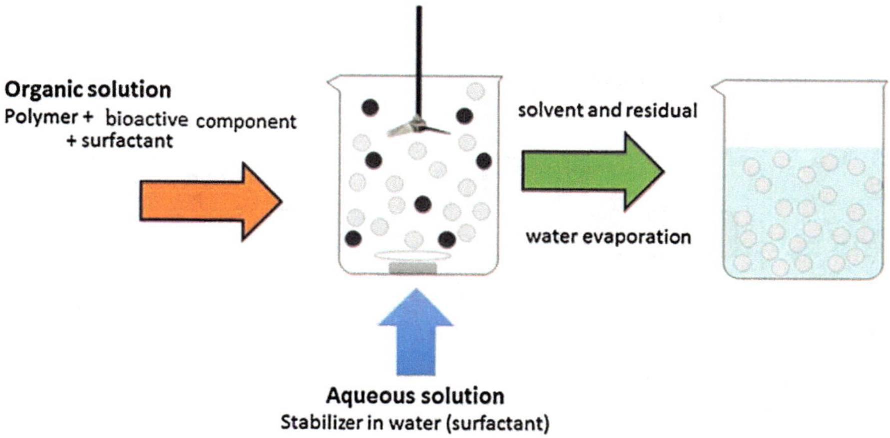

Fig. 7.6 Schematic representation of the nanoprecipitation technique

with high encapsulation efficiency, high reproducibility, narrow size distribution, less physical stress and easy scaling-up procedure. The aqueous phase and the organic phase, including biopolymer, carrier oil and organic solvent, are prepared separately, and an emulsion is then fabricated using a mechanical shearing method. When excess water is added to solution, fast elimination of the organic solvent and the separation of the biopolymer and the oil, particle size decrease, biopolymer precipitate, and finally nanocarrier formation occurs. Finally, the organic solvent can be evaporated under reduced pressure by a rotary evaporator. The organic solvent must have high solubility in both oil and polymer and partial solubility in water so that diffusion after dilution is possible (Fig. 7.7) (Anarjan et al. 2016). The shear rate and temperature used in the emulsification process, the chemical compositions and amount of organic phase and emulsifier, the polymer concentration and the oil-to-polymer ratio play an important role in the nanocapsules' size. This method provides an easy procedure for nanoencapsulation of different food bioactive compounds; however, since some residual organic solvents might remain in the final product, it should be aware of their potential toxic effects (Fathi et al. 2014).

Furthermore, use of multiple emulsions, including a complex of biopolymers, increase the stability and encapsulation efficiency of the bioactive compounds. Different vitamins can be encapsulated and transmitted via nanoemulsions, for example nanoencapsulation of vitamins and bioactive compounds derivative covered with lecithin as an edible encapsulant (Katouzian and Jafari 2016).

7.4.3 Coacervation Method

Coacervation encapsulation includes the phase separation of one or more hydrocolloids from solution because of macromolecules desolvation and the subsequent deposition of the newly formed coacervate phase in the same reaction medium.

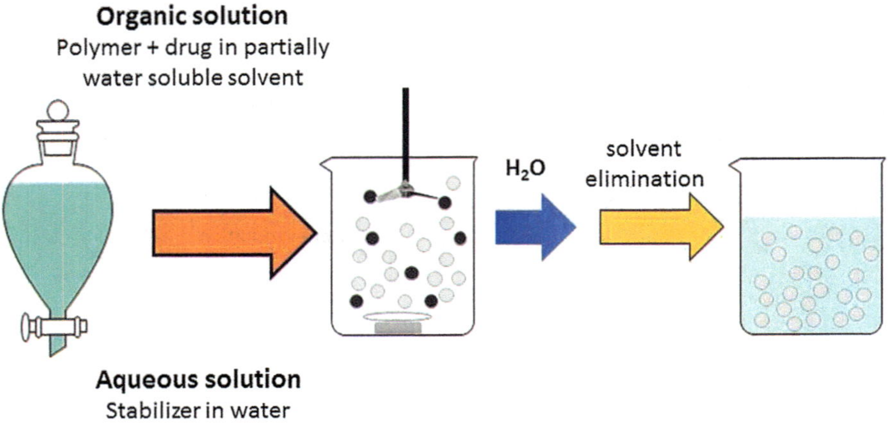

Fig. 7.7 Schematic representation of the emulsification/solvent diffusion technique

This method has been used for both non-polar and polar bioactive molecules (Abbas et al. 2012). Coacervation is one of the most commonly used methods for production of carbohydrate-based delivery systems. The electrostatic attraction between oppositely charged molecules causes driving force for this method. This force may be induced between a charged bioactive component and an oppositely charged carbohydrate. Alternatively, a bioactive may be trapped within a particle formed by electrostatic complexation of positively charged, such as chitosan, and a negatively charged, such as pectin and alginate biopolymers. The functional performance of the produced nanocapsules depends upon the chemical nature and the surface characteristics of the biopolymeric shell that some of depends parameters are written in following (Fathi et al. 2012, 2014).

Butstraen et al. showed that the higher the surface charge (a pH dependent parameter) had, the better the performance of the nanoencapsulation surface charge (Butstraen and Salaün 2014). Piacentini and coworkers discussed the rate of agitation that aggregation might be happened in a very low or very high stirring rate. Also, they established that the rate of lower dropwise addition of polymer solution caused better the performance and higher solubilities of bioactive and biopolymer in solution can cause better performance (Fathi et al. 2014).

Sumithra and coworkers studied the herbal extract enclosed bovine serum albumin protein by coacervation process followed by cross-linking with glutaraldehyde (Sumithra and Raaja 2012).

Zhou et al. studied encapsulation of hepatocytes' living cells via polyelectrolyte complex coacervation between the cationic methylated collagen and anionic terpolymer of hydroxyethyl methacrylate, methyl methacrylate and methylacrylic acid (Jaworek 2008).

Polymeric nanoparticles are prepared by using biodegradable hydrophilic polymers such as chitosan, gelatin and sodium alginate. Calvo et al. (1997a, b) modified a method for preparing hydrophilic chitosan nanoparticles by ionic gelation. The formation of coacervates from cationic chitosan and anionic alginate through electrostatic complexation is depicted in Fig. 7.8 (Fathi et al. 2014).

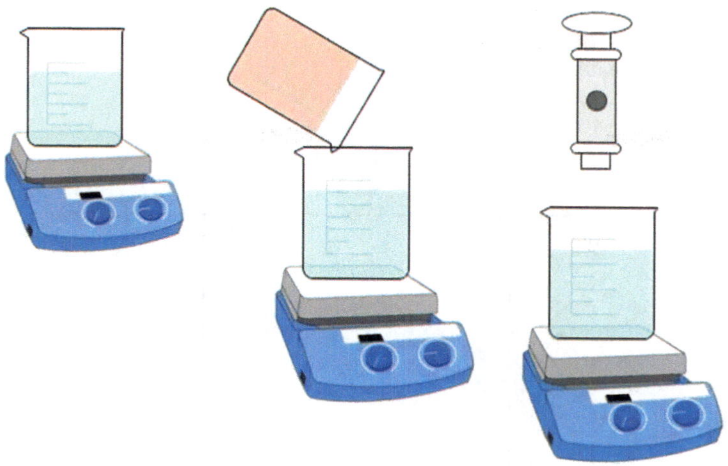

Fig. 7.8 Schematic representation of the coacervation technique

7.4.4 Spray Drying Method

Spray drying is one of the most commonly used techniques for encapsulation. It is fast, relatively reproducible, flexible, economical and allows for continuous operation making it an effective drying method to encapsulate and increase the stability of sensitive bioactive compounds. The principle of this technique is based on dissolving or dispersing the bioactive compounds in a solution of biopolymer. Then the dispersion is atomized in a heated air chamber (Fig. 7.9), which rapidly removes the solvent and produces a dried particle consisting of the bioactive compounds inserted in a porous wall material. Wall material used for spray drying should have high water solubility, low viscosity at higher concentrations and good emulsification. Gum acacia, maltodextrins, hydrophobically modified starch, and their mixtures are commonly used for wall materials (Abbas et al. 2012; Esfanjani and Jafari 2016).

So a challenge for drying engineers is to design a process that reaches efficient encapsulations and good product potentials without damage and loss to sensitive bioactive compounds (Fathi et al. 2014; Jarunglumlert and Nakagawa 2013). When the maximum possible amount of core material is encapsulated inside the particles, efficient encapsulations are obtained. A great deal of research has been developed to study the effects of operating conditions during spray drying to improve encapsulation efficiency (Jarunglumlert and Nakagawa 2013).

The size of produced particles and the efficiency of encapsulation depends on a number of factors such as the properties of the materials used, for example core and wall material properties; solution viscosity; spray dryer features, for example atomizer type, flow rates, inlet and outlet temperatures, feed temperature, solid matter concentration and humidity (Fathi et al. 2014; Jarunglumlert and Nakagawa 2013).

Some authors (Aghbashlo et al. 2012; Carneiro et al. 2013; Jafari et al. 2008) established that lower feed flow rates create small particle sizes and higher efficiency.

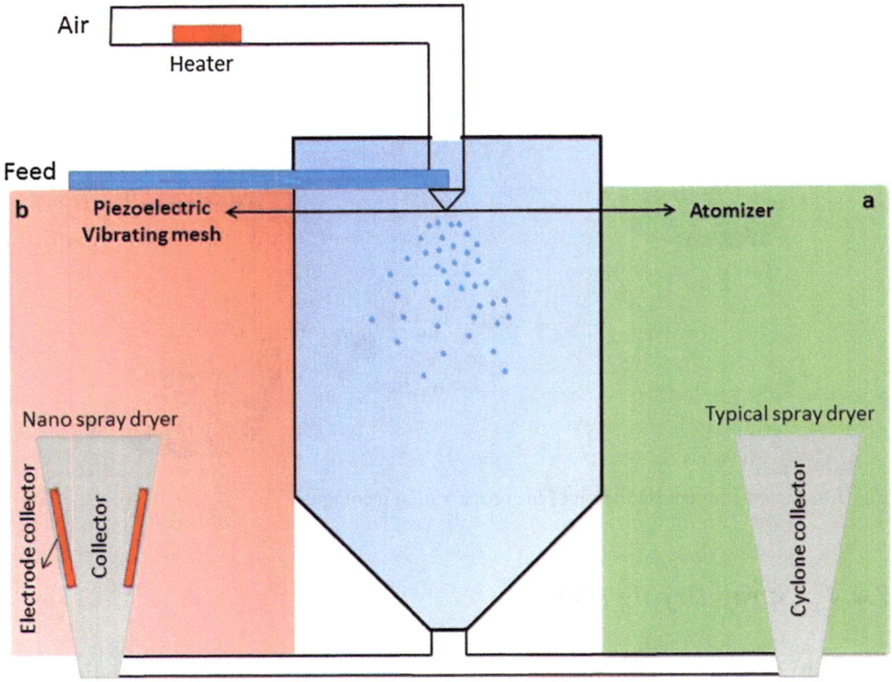

Fig. 7.9 Schematic representation of spray dryer technique

Higher air flow rate causes a larger size and a moderate air flow rate leads to high efficiency. Higher inlet temperatures cause larger particle sizes and higher efficiency due to fast crust formation.

Because of high temperature and gelation, the possibility of solution used in this process, application of volatile or thermo-sensitive bioactive compounds and some carbohydrates, such as starch, might be limited. On the other hand, thermostable carbohydrate biopolymers, such as cyclodextrins or modified materials, such as hydroxypropyl cellulose, are suitable for spray drying at high temperatures. Common spray dryers use rotary atomizers and pressure nozzles or two-fluid nozzles to produce spray droplets. The submicron particles are produced with these dryers (Fathi et al. 2014).

To produce nanoparticles, a new generation of spray-dryers (Nano Spray Dryer) was recently used where nanoparticles are obtained by effective fluid breakdown joined with highly efficient particle collectors. This technology uses a vibrating mesh for very narrow droplet size generation. Dried particles are then collected by an electrostatic particle collector, which is size independent in contrast to cyclones (Fig. 7.9). The smaller capsules are produced by direct spray drying than by coacervation-spray drying (Fathi et al. 2014; Ghayempour and Montazer 2016).

The efficiency of this technique is high for producing nano sized carriers in laboratory scale. Nano scale carriers have been created using this technology for entrapment of bovine serum albumin and some pharmaceutical products (Fathi et al. 2014).

7.4.5 Electrospinning and Electrospray Methods

Electrospinning has expanded widespread interest as a method for fabrication of continuous nanofibers which could be used for the loading and delivering of bioactive compounds. This technique is a simple, one-step processes for the production of micro and nanoencapsulates in the form of dried particles. Electrospinning uses electrostatic forces to generate a wide range of biocompatible, biodegradable, food grade, conducting polymeric substances as wall materials for encapsulation of bioactive compounds (Bhushani and Anandharamakrishnan 2014). For fabricating nanoencapsulate the polymer solution contains the following processes:

1. a capillary through which the polymer solution is to be electrospun is pumped;
2. a high voltage source with positive or negative polarity, which generates a charge in the polymer solution. This potential leads to the formation of repulsive interactions between the same charges in the solution and the attractive forces between the oppositely charged solution and collector and consequently elongating the pendant drop;
3. a grounded collector, which is brought into contact with the counter electrode (Fig. 7.10) (Fathi et al. 2014).

In electrospinning, a spinneret produces a droplet at the spinneret exit from a polymer solution. The electric charges will collect on the surface of the droplet using an electrical field. Next, the shape of cone droplets will be formed by the electric field (Katouzian and Jafari 2016).

Nanoencapsulation formation and structure depend on different parameters such as used voltage, solution flow rate, capillary–collector distance, solution properties, bioactive compounds concentration and solvent volatility. Higher voltage initially causes the formation of thinner and then thicker shells, longer distance, lower concentration, higher flow rate, higher conductivity and higher volatility leads to a thinner shell of encapsulation (Fathi et al. 2012).

Also, this method was used for encapsulation and delivery of bifidobacteria, vanillin, NiO/TiO_2, lysozyme, and epigallocatechin-3-gallate. The starch fibers containing amylose inclusion complexes produced by electrospinning (Zhu 2017).

Stijnman et al. (2011) has studied electrospinning of food-grade polysaccharides from plants such as galactomannans, glucomannans, starch and methylcellulose, seaweeds such as carrageenans and alginate and micro-organisms such as pullulan and dextran.

The production of gelatin nanofibers was produced by electrospinning, which can be used as a more effective thickening agent in lower amounts compared to bulk gelatin. It has also been noted that gelatin nanofibers can be applied in the stabilization of food emulsions. According to Neo et al. (2013), zein nanofibers could be loaded with antioxidants, such as gallic acids (Ramachandraiah et al. 2015).

Electrospray is a new nanoencapsulation technique which it's basic is close to electrospinning; however, instead of production of nanofibers, nanoparticles are formed (Neo et al. 2013). In this technique high voltage causes the electrostatic force that atomizes liquid into fine droplets. The solvent evaporation is completed

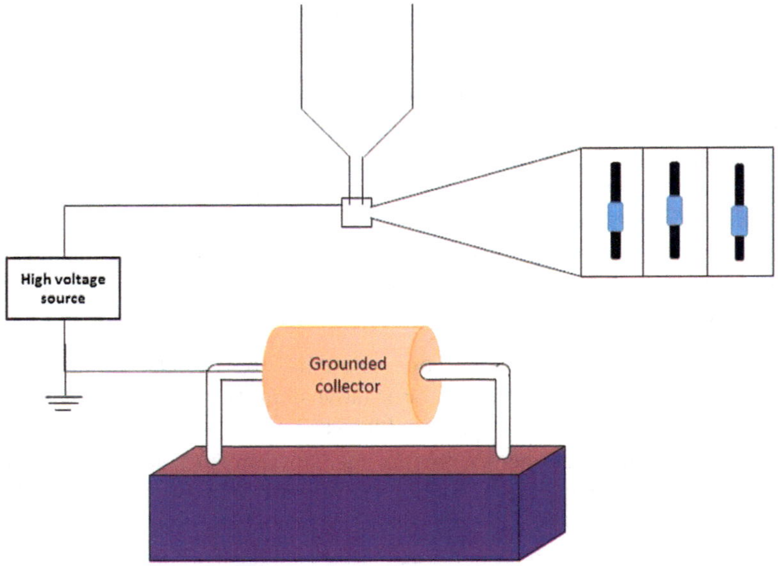

Fig. 7.10 Schematic representation of the electrospinning technique

during the flight of the droplets to the ground electrode. The high encapsulation efficiency and possibility of the production in one step are important properties of electrospray technique. Thinner fibers or smaller particles with a high surface charge are produced from a solution of carbohydrate biopolymers using electrospinning, electrospray and coacervation methods. Properties of biopolymers are improved by chemical grafting or their combination with highly charged materials for carrier production that increase their surface charge (Fathi et al. 2014).

7.4.6 Solvent Evaporation

In the solvent evaporation method, polymer solutions are prepared in volatile organic solvents (Nagavarma et al. 2012). Solvent evaporation as the main dispersion method is based on O/W (oil-in-water) emulsions by using a surfactant and solvent, then the solvent is removed by evaporation (Fig. 7.11). The solvent evaporation method can be used for the nanoencapsulation of phenolics such as curcumin in polymeric nanoparticles (Esfanjani and Jafari 2016). Furthermore, solvent evaporation remains to be an exclusive approach to encapsulating lipophilic vitamins (Katouzian and Jafari 2016).

In this method, the polymers are dissolved in organic solvent and poured into continuously stirring aqueous phase with or without emulsifier/stabilizer and sonicated. Most of the bioactive compounds are denatured after the sonication. Thus,

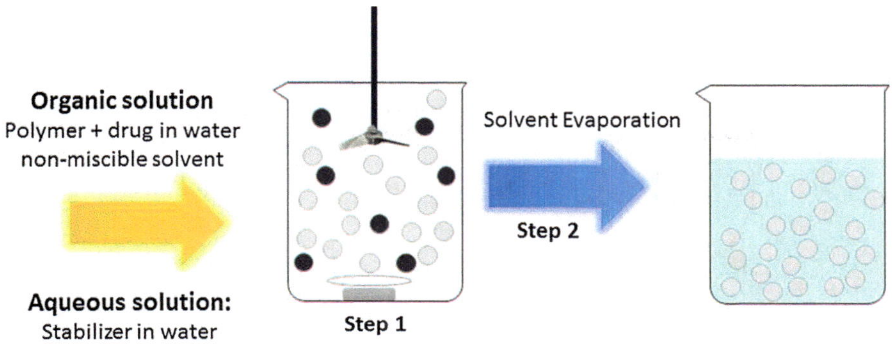

Fig. 7.11 Schematic representation of the solvent-evaporation technique

sonication is done at a slow rate and low temperature that is effective in retaining the structure of bioactive compounds and drugs (Sohani 2013).

7.4.7 Other Methods

There are different methods for the production of bioactive compound nanoencapsulation. In this section and Fig. 7.12, some of the other important encapsulation techniques are summarized (Ghayempour and Montazer 2016).

7.5 Nanostructured Delivery Systems Applied in Encapsulation and Controlled Release

Recently, the study of the behavior of nanosystems has attracted attention as carriers of biological materials and in the understanding control of those systems for functions such as delivery, control release and protection for food and pharmaceutical applications (Cerqueira et al. 2014). It is necessary to employ strategies to enhance the delivery, release and retention of bioactive compounds in cells and tissues of body. The use of encapsulation technologies is a possible method to own all these values (Mozafari et al. 2006).

The delivery systems can be classified into four categories of technologies:

1. Solid and powdered delivery technologies such as microspheres, microcapsules, nanospheres, and micro- or nanoparticles
2. Liquid micro and nano vehicles such as emulsions, miniemulsions, nanoemulsions and microemulsions
3. Organogels and aqueous gels
4. Lyotropic liquid crystals.

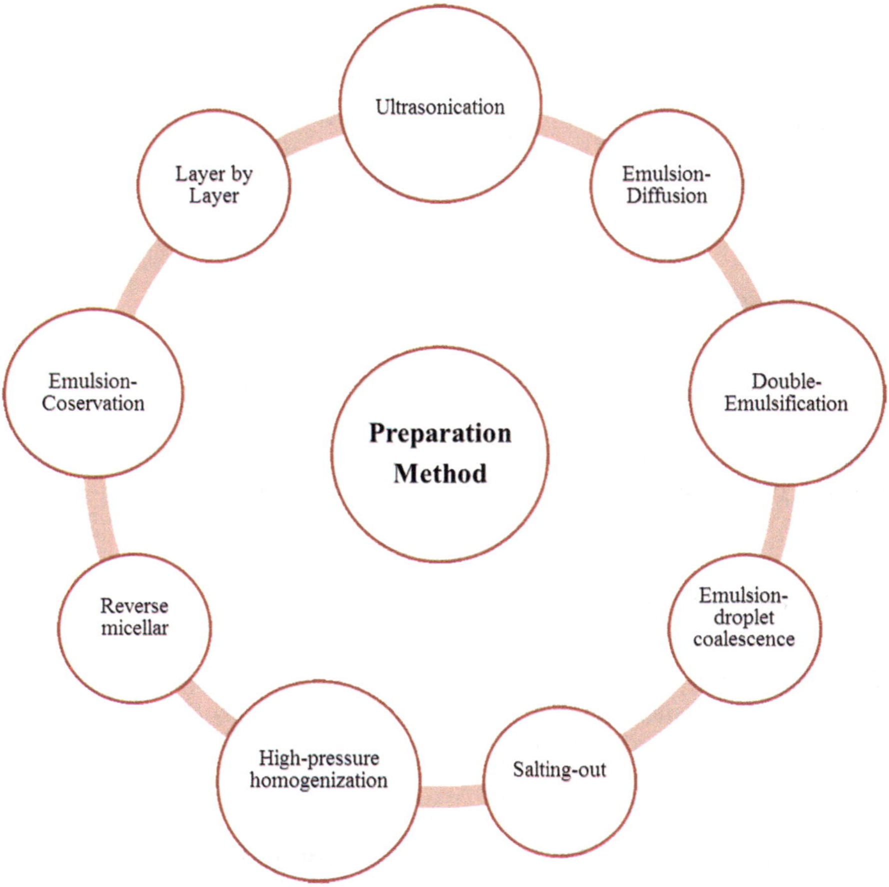

Fig. 7.12 Some common methods for nanoencapsulation

Nanoencapsulated compounds have been used initially for the design of drug delivery systems. In the near future, it is assumed that nanoencapsulated bioactive compounds will be generally used in the food, beverage, cosmetic and drug delivery systems (Katouzian and Jafari 2016).

In developing these systems, it is essential to consider the end application, the limitations and benefits offered by encapsulation over the presentation of a neat bioactive (Augustin et al. 2013). This material for drug delivery application should be non-toxic, safe and biodegradable in the human body which needs toxicological safety evaluation before being used for human consumption. The common testing of nanoparticles is not suitable for toxicity tests of nanoencapsulated compounds. The effectiveness of nanoencapsulated compounds must also be tested to ensure that the process is actually improving the characteristic of non-encapsulated compounds (Gutiérrez et al. 2013; Luengo et al. 2006).

Nanoscale cores of material are encapsulated with an outer shell for controlled release systems. The core must be insoluble under some conditions, such as low pH, and soluble under other conditions at which controlled release is happening.

Interaction of the polymeric matrix with different environmental parameters, such as temperature, enzymes, pH and ionic strength, causes degradation. This allows control of the amount of bioactive compounds released over time. The bioactive component release profile depends on polymeric degradation rates and changes in the diffusion process. Also the release rate usually depends on the thickness of the encapsulating shell and the material used in the coating. A longer release time lasted for thicker shells.

Several studies published in the literature focus on the release of bioactive compounds, a model component for hydrophobic materials. It is possible to design a varied targeted drug delivery system that releases the bioactive compound to a specific site of action, such as mouth, stomach, small intestine, large intestine or colon, and to protect of the bioactive compound against chemical and enzymatic degradation and potentially enhance the selective uptake of these particles (Amjadi et al. 2013; Cerqueira et al. 2014; Dang et al. 2015; Jung et al. 2016).

These systems can be used as controlled drug delivery devices that have numerous advantages compared with conventional dosage forms, including improved efficacy, reduced toxicity and improved patient convenience (Dupeyrón et al. 2005).

Bioactive compounds released before and after encapsulation are observed with a UV–Vis spectrometer (Ai et al. 2003). The release profile and optimal targeted delivery of bioactive compounds can be predicted using novel computational and numerical methods like Monte-Carlo simulations (Katouzian and Jafari 2016).

7.6 The Future of Controlled Release Systems

In the delivery systems, bioactive compounds released at the tissue site are done by any one of the three general physico-chemical mechanisms:

Mechanism 1: By the swelling of the polymer bioactive compounds and hydration followed so release through diffusion.

Mechanism 2: By an enzymatic reaction resulting in break or cleavage or degradation of the polymer at site of delivery by releasing the bioactive compounds from the entrapped inner core.

Mechanism 3: Dissociation of the bioactive compounds from the polymer and its de-adsorption/release from the swelled (Nagavarma et al. 2012).

It is expected that the future will bring many new products that are based on loading various mixtures of additives, flavors, nutrients and bioactive compounds. It is a very big challenge to modify these systems with insoluble bioactive compounds in water. In the solubilization and the bioavailability technology, it will be very important to control, trigger and slow release bioactive compounds and nutritional additives into the final product that we use in our body. In the next step, scientists must focus on building the carrier vehicles with certain characteristics or to embed within them that which will control the release of the bioactive compounds from the product into our body or across blood barrier. It seems that they have a long way to develop. However, the pharmaceutical industry has progressed in the field of controlling the release of bioactive compounds, but these works are mainly expensive. They hope that many of the controlled release features that are impossible now will be the challenge in the future (Garti 2008).

Recently, there has been a considerable interest in the development of nano-scale delivery systems for bioactive compounds because of improved bioavailability and stability. Nanoencapsulation provides final product functionality, including controlled release of the bioactive compounds, which is expected to be maintained during storage. So studies on nanoencapsulation and delivery of bioactive compounds by different methods have been reviewed in the present work. Although a number of different types of material delivery systems have been discussed, many of these studies in food industries involved nanoencapsulation to protect the antioxidants from degradation during process and storage.

References

Abbas S, Da Wei C, Hayat K, Xiaoming Z. Ascorbic acid: microencapsulation techniques and trends—a review. Food Rev Int. 2012;28(4):343–74.

Aghbashlo M, Mobli H, Madadlou A, Rafiee S. Integrated optimization of fish oil microencapsulation process by spray drying. J Microencapsul. 2012;29(8):790–804.

Ai H, Jones SA, de Villiers MM, Lvov YM. Nano-encapsulation of furosemide microcrystals for controlled drug release. J Control Release. 2003;86(1):59–68.

Amjadi I, Rabiee M, Hosseini M, Sefidkon F, Mozafari M. Nanoencapsulation of hypericum perforatum and doxorubicin anticancer agents in PLGA nanoparticles through double emulsion technique. Micro Nano Lett. 2013;8(5):243–7.

Anandharamakrishnan C. Techniques for nanoencapsulation of food ingredients. Berlin: Springer; 2014.

Anarjan N, Ghaz Jahanian MA, Sayyar Z, Mohamadlou M, Najafi S, Pazhohnia Z, Jaberi N. Production of nanodispersions by solvent-displasment technique. In: Jafarizadeh-Malmiri H, Berenjian A, editors. High value processing technologies. Hauppauge: Nova Science Publishers; 2016.

Augustin MA, Sanguansri L, Lockett T. Nano-and micro-encapsulated systems for enhancing the delivery of resveratrol. Ann N Y Acad Sci. 2013;1290(1):107–12.

Bhushani JA, Anandharamakrishnan C. Electrospinning and electrospraying techniques: potential food based applications. Trend Food Sci Technol. 2014;38(1):21–33.

Butstraen C, Salaün F. Preparation of microcapsules by complex coacervation of gum Arabic and chitosan. Carbohydr Polym. 2014;99:608–16.

Calvo P, Remuñan-López C, Vila-Jato JL, Alonso MJ. Chitosan and chitosan/ethylene oxide-propylene oxide block copolymer nanoparticles as novel carriers for proteins and vaccines. Pharm Res. 1997a;14(10):1431–6.

Calvo P, Remuñan-López C, Vila-Jato JL, Alonso M. Novel hydrophilic chitosan-polyethylene oxide nanoparticles as protein carriers. J Appl Polym Sci. 1997b;63(1):125–32.

Carneiro HC, Tonon RV, Grosso CR, Hubinger MD. Encapsulation efficiency and oxidative stability of flaxseed oil microencapsulated by spray drying using different combinations of wall materials. J Food Eng. 2013;115(4):443–51.

Cerqueira MA, Pinheiro AC, Silva HD, Ramos PE, Azevedo MA, Flores-López ML, Rivera MC, Bourbon AI, Ramos OL, Vicente AA. Design of bio-nanosystems for oral delivery of functional compounds. Food Eng Rev. 2014;6(1–2):1–19.
Dang S, Gupta S, Bansal R, Ali J, Gabrani R. Nano-encapsulation of a natural polyphenol, green tea catechins: way to preserve its antioxidative potential. In: Free radicals in human health and disease. Springer; 2015. p. 397–415.
Dupeyrón D, González M, Sáez V, Ramón J, Rieumont J. Nano-encapsulation of protein using an enteric polymer as carrier. In: IEE Proceedings-Nanobiotechnol, vol. 152. IET; 2005; p. 165–168.
Esfanjani AF, Jafari SM. Biopolymer nano-particles and natural nano-carriers for nano-encapsulation of phenolic compounds. Colloids Surf B: Biointerfaces. 2016;146:532–43.
Fathi M, Mozafari M-R, Mohebbi M. Nanoencapsulation of food ingredients using lipid based delivery systems. Trend Food Sci Technol. 2012;23(1):13–27.
Fathi M, Martín Á, McClements DJ. Nanoencapsulation of food ingredients using carbohydrate based delivery systems. Trend Food Sci Technol. 2014;39(1):18–39.
Garti N. Delivery and controlled release of bioactives in foods and nutraceuticals. Amsterdam: Elsevier; 2008.
Ghayempour S, Montazer M. Micro/nanoencapsulation of essential oils and fragrances: focus on perfumed, antimicrobial, mosquito-repellent and medical textiles. J Microencapsul. 2016;33(6):497–510.
Gutiérrez FJ, Albillos SM, Casas-Sanz E, Cruz Z, García-Estrada C, García-Guerra A, García-Reverter J, García-Suárez M, Gatón P, González-Ferrero C. Methods for the nanoencapsulation of β-carotene in the food sector. Trend Food Sci Technol. 2013;32(2):73–83.
Hădărugă N, Hădărugă D, Bandur G, Riviş A, Pârvu D, Lupea A. Protection and controlled release of fatty acids and essential oils by nanoencapsulation in cyclodextrins (a review). J Agroalimentary Process Technol. 2008;14(2):381–9.
Jafari SM, Assadpoor E, Bhandari B, He Y. Nano-particle encapsulation of fish oil by spray drying. Food Res Int. 2008;41(2):172–83.
Jarunglumlert T, Nakagawa K. Spray drying of casein aggregates loaded with β-carotene: influences of acidic conditions and storage time on surface structure and encapsulation efficiencies. Dry Technol. 2013;31(13–14):1459–65.
Jaworek A. Electrostatic micro-and nanoencapsulation and electroemulsification: a brief review. J Microencapsul. 2008;25(7):443–68.
Jung EY, Hong KB, Son HS, Suh HJ, Park Y. Effect of layer-by-layer (LbL) encapsulation of nano-emulsified fish oil on their digestibility ex vivo and skin permeability in vitro. Prev Nutr Food Sci. 2016;21(2):85.
Katouzian I, Jafari SM. Nano-encapsulation as a promising approach for targeted delivery and controlled release of vitamins. Trend Food Sci Technol. 2016;53:34–48.
Livney YD. Nanostructured delivery systems in food: latest developments and potential future directions. Curr Opinion Food Sci. 2015;3:125–35.
Luengo J, Weiss B, Schneider M, Ehlers A, Stracke F, König K, Kostka K-H, Lehr C-M, Schaefer U. Influence of nanoencapsulation on human skin transport of flufenamic acid. Skin Pharm Physiol. 2006;19(4):190–7.
Luo Y, Zhou X. Nanoencapsulation of a hydrophobic compound by a miniemulsion polymerization process. J Polym Sci Part A Polym Chem. 2004;42(9):2145–54.
Mozafari MR, Flanagan J, Matia-Merino L, Awati A, Omri A, Suntres ZE, Singh H. Recent trends in the lipid-based nanoencapsulation of antioxidants and their role in foods. J Sci Food Agric. 2006;86(13):2038–45.
Nagavarma B, Yadav HK, Ayaz A, Vasudha L, Shivakumar H. Different techniques for preparation of polymeric nanoparticles-a review. Asian J Pharm Clin Res. 2012;5(3):16–23.
Neo YP, Ray S, Jin J, Gizdavic-Nikolaidis M, Nieuwoudt MK, Liu D, Quek SY. Encapsulation of food grade antioxidant in natural biopolymer by electrospinning technique: a physicochemical study based on zein-gallic acid. Food Chem. 2013;136(2):1013–21.

Noronha CM, Granada AF, de Carvalho SM, Lino RC, de OB Maciel MV, Barreto PLM. Optimization of α-tocopherol loaded nanocapsules by the nanoprecipitation method. Ind Crops Prod. 2013;50:896–903.

Nuruzzaman M, Rahman MM, Liu Y, Naidu R. Nanoencapsulation, nano-guard for pesticides: a new window for safe application. J Agric Food Chem. 2016;64(7):1447–83.

O'Connor G, Gleeson LE, Fagan-Murphy A, Cryan S-A, O'Sullivan MP, Keane J. Sharpening nature's tools for efficient tuberculosis control: a review of the potential role and development of host-directed therapies and strategies for targeted respiratory delivery. Adv Drug Deliv Rev. 2016;102:33–54.

Pandey R, Ahmad Z, Sharma S, Khuller G. Nano-encapsulation of azole antifungals: potential applications to improve oral drug delivery. Int J Pharm. 2005;301(1):268–76.

Quintanilla-Carvajal MX, Camacho-Díaz BH, Meraz-Torres LS, Chanona-Pérez JJ, Alamilla-Beltrán L, Jimenéz-Aparicio A, Gutiérrez-López GF. Nanoencapsulation: a new trend in food engineering processing. Food Eng Rev. 2010;2(1):39–50.

Ramachandraiah K, Han SG, Chin KB. Nanotechnology in meat processing and packaging: potential applications—a review. Asian-Australasian J Anim Sci. 2015;28(2):290.

Reis CP, Neufeld RJ, Ribeiro AJ, Veiga F. Nanoencapsulation I. Methods for preparation of drug-loaded polymeric nanoparticles. Nanomedicine. 2006a;2(1):8–21.

Reis CP, Neufeld RJ, Ribeiro AJ, Veiga F. Nanoencapsulation II. Biomedical applications and current status of peptide and protein nanoparticulate delivery systems. Nanomedicine. 2006b;2(2):53–65.

Sohani OR. Nanoencapsulation system for delivery of protein and peptide—a review. J Biomed Pharm Res. 2013;2(2):58–64.

Stijnman AC, Bondar I, Tromp RH. Electrospinning of food-grade polysaccharides. Food Hydrocoll. 2011;25(5):1393–8.

Sumithra M, Raaja NV. Micro-encapsulation and nano-encapsulation of denim fabrics with herbal extracts. Indian J Fibre Text Res. 2012;37(4):321–5.

Xiao J, Cao Y, Huang Q. Edible nanoencapsulation vehicles for oral delivery of phytochemicals: a perspective paper. J Agric Food Chem. 2017;65(32):6727–35.

Xing F, Cheng G, Yi K, Ma L. Nanoencapsulation of capsaicin by complex coacervation of gelatin, acacia, and tannins. J Appl Polym Sci. 2005;96(6):2225–9.

Yadav SC, Kumari A, Yadav R. Development of peptide and protein nanotherapeutics by nanoencapsulation and nanobioconjugation. Peptides. 2011;32(1):173–87.

Zhu F. Encapsulation and delivery of food ingredients using starch based systems. Food Chem. 2017;229:542–52.

Chapter 8
Potential Hazards of Nanoparticles

8.1 Introduction

Recent developments in the design of advanced materials have furthered interest in the commercialization of new technologies. So the increased production of nanomaterials has increased concerns about their effects on human and environmental health. The evidence for health risks of nanoparticles has been demonstrated over the last decade, yet it is unclear if metal nanoparticles cause effects directly or indirectly. This chapter gives a brief review on the toxicology pathways, recommendations and methods for screening hazard testing of nanoparticles.

Nanotechnology plays an important role in the improvement of the food and drug industry in the last few years and will have an enormous impact on life sciences, including drug delivery, food, pharmacy, engineering and the production of biomaterials (Forbe et al. 2011).

The reason why nanoparticles are attractive for such purposes is based on their important and unique features, such as their small size and special surface area, and which special surface area is much larger than that of other particles and materials (Fig. 8.1). These features can cause an increase in the risk of fire and explosion compared to other particles. Therefore, nanoparticles have a large surface which might be chemically more reactive to bind, adsorb and carry other compounds such as drugs, probes and proteins compared to their fine analogues (Borm and Kreyling 2004; Oberdörster 2010).

Along with the developments in nanotechnology and new products, these materials can be harmful for human health and the environment. In nanometer scale, material properties change; along with these changes, the prediction, identification, evaluation and control of the health, safety and environmental risks of nanomaterials are also challenged (Oberdörster 2010).

Risk estimation has been supposed of as the evaluation of what can go wrong, how this is likely to occur. Risk estimation has been the guiding standard for the

H. Jafarizadeh-Malmiri et al., *Nanobiotechnology in Food: Concepts, Applications and Perspectives*, https://doi.org/10.1007/978-3-030-05846-3_8

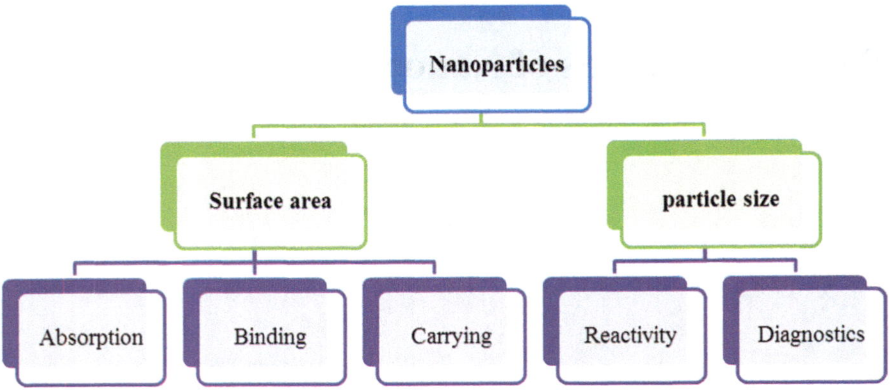

Fig. 8.1 Properties of nanoparticles for their potential biological effects

evaluation of environmental and product risks. In the case of chemicals and nanomaterials, risk estimation has relied on detailed, experimental data for exposure and hazard. Objective driven approaches, referred to here as top-down and bottom-up methods, rely on the achievement of information and synthesis from result makers to drive actions. Top-down methods can improve the risk estimation process by mixing technical information and expert results on an emerging technology with human values (Fadel et al. 2015).

In all studies on the effects of nanomaterials, two groups of different nanostructures should be considered:

(a) Fixed nanoparticles, nanocomposites, nanostructured surfaces and nanoparticle components.
(b) Free nanoparticles; these nanoparticles can create a complex combination of other nanoparticles from a specific element that is coated with another material.

Fixed nanoparticles are used in many current applications that are not naturally dispersive. New applications include coatings, textiles, ceramics, membranes, composite materials, glass products, prosthetic implants, anti-static packaging, cutting tools, industrial catalysts, a variety of electric and electronic devices including displays, batteries and fuel cells (Salata 2004; Royal Society and Royal Academy of Engineering 2004).

Other uses of nanoparticles include drugs, biodegradable materials for biomedical, personal care products, such as cosmetics, quantum dots and some pilot applications in environmental remediation (Fig. 8.2). Also potential exposure to manufacture nanoparticles may increase naturally in the future (Assa et al. 2015; Guadagnini et al. 2015; Vauthier et al. 2003).

Researchers now focus on hazards of nanoparticles that are currently used in the production of products and on the subject of what can be done to limit the related risks. However, research on hazards of nanoparticles is still limited (Reijnders 2006).

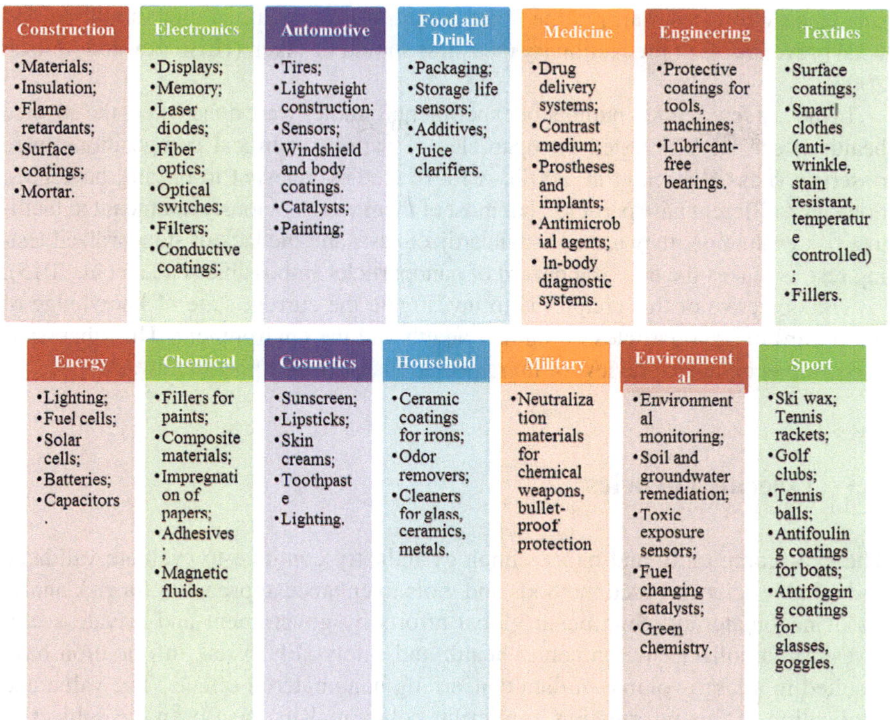

Fig. 8.2 overview of areas of applications of nanotechnology

8.2 Hazard and Risk Definition

Nanotoxicology, an important subcategory of nanotechnology, has been defined as science of nanostructures that has an effect on living organisms (Borm et al. 2006). The term hazard is defined in different forms. In this chapter the term hazard is used following the definition of the United States Environmental Protection Agency (USEPA), which defines hazard as the inherent toxicity of a compound. According to this definition, if a chemical material has the toxicity property, it is hazardous. Any exposure to a hazardous material may lead to adverse health effects in humans. USEPA defines risk with respect to the above definition of hazard as a measure of the probability of damage to life, health, property and the environment. According to this definition, because the probability of exposure to a hazardous material is high and the consequences cause risk for the health or environment it is therefore important to estimate risk (Kavlock and Dix 2010; Warheit 2004). Usually known risks and potential risks are two categories of risk that are distinguished in literature. When the causal relationship between a cause and damage is established, prevention is possible. In case of potential risks, it is unclear whether there is a danger, how significant the damage can be or what is the probability of its occurrence. To evaluate the risks of hazardous agents is very important. The likelihood that a hazardous substance will cause harm is the determinant of how cautious one should be and what preventative or precautionary measures should be taken (Hristozov and Malsch 2009).

In the last few years, a number of experimental studies were done where the adverse health effects of 965 tested nanoparticles of various chemical compositions were observed (Foss Hansen et al. 2007). A lot of studies, relevant to hazard, have been done with different nanoparticles, but most of them were obviously not meant to facilitate risk evaluation; they use non-standardized tests and the lack of standardized testing results makes the univocal hazard of nanoparticles impossible (Fadel et al. 2015).

The objective of this chapter is to investigate the current state of knowledge of the hazard of nanoparticles on human health and the environment. This chapter is based on an extensive review of literature published in the past to now.

8.3 Overall Theories

The U.S. community and nanotechnology industry continue to evaluate validated and reliable science-based methods and tools to enhance approaches for risk analysis of nanomaterials. Significant global efforts by government and private sector investors to collect environmental health and safety (EHS) risk information have resulted in a large volume of data concerning nanomaterial effects. The value and application of this information to relevant policy makers have been the subject of multiple national and international efforts, including four workshops organized by the National Nanotechnology Initiative (NNI)—a strategy exposing potential nanotechnology risks—from 2009 to 2010 (Roco 2007). Various investor communities, such as industry, workers, consumers and non-government organizations, should confirm that novel nanomaterials are safe for human health and the environment. Collected EHS information by government decision makers and private sector investors have helped to create data concerning nanomaterial effects such as information about the toxicity of a compound and determining limits for exposure and uptake into the body. Such research must be followed by the modified methods for characterizing and quantifying exposure to humans and the environment. To consider the best path in 2009 for nanotechnology, NNI agencies studied the 2008 EHS Research Strategy and data in EHS. In 2008 the Nanotechnology Environmental and Health Implications (NEHI) Working Group of the NNI published a document titled *Strategy for Nanotechnology-Related Environmental, Health, and Safety Research*. The EHS Research Strategy document represented the conclusion of a comprehensive effort led by the Nanotechnology Environmental and Health Implications (NEHI) Working Group under the National Science, Nanoscale Science, Engineering, and Technology Subcommittee to provide guidance to the NNI Federal agencies producing scientific information for risk management (Howard 2013; Roco 2007; Schmidt 2009).

To share the latest information, newest developments and research gaps in the nanotechnology EHS field, workshops were organized. Overall, the knowledge collected from the workshops were critical to the development of the NNI EHS

Research Strategy and established standards. This strategy document identified important data needed in the areas of nanomaterial measurement organization, risk valuation and management methods, human health and the environment.

The 2013 NNI workshops were designed to facilitate discussion among various investors of approaches, equipment and methods used to measure, manage and communicate the potential risks of nanomaterials and nanotechnology-based products (Fig. 8.3) (Howard 2013).

Associates discussed the importance of providing sufficient information on nanomaterial products in safety in order to protect human health and the environment against damage of nanomaterials. The challenges of nanotechnology risk management suggested the creation of information that includes guidance documents to aid the small business community in making informed decisions when entering the nanotechnology market (Maynard et al. 2006).

Several requirements were discussed such as techniques to measure particles in vivo, the release of particles into the environment and the toxicity potential of nanomaterials. Particle uniformity might also influence the kinetics and toxicity of nanomaterials in unknown ways. Universal attempts have developed strategies for improving the data quality for characterizing nanomaterials, toxicological studies and validated methods for characterizing and quantifying nanomaterials in experimental media, tissues and cells. Also, attempts in the private sector have concentrated on the development of new methods to estimate 'commercial' risk and exposure control measures, and the scientific and regulatory information to evaluate toxicity (Schmidt 2009).

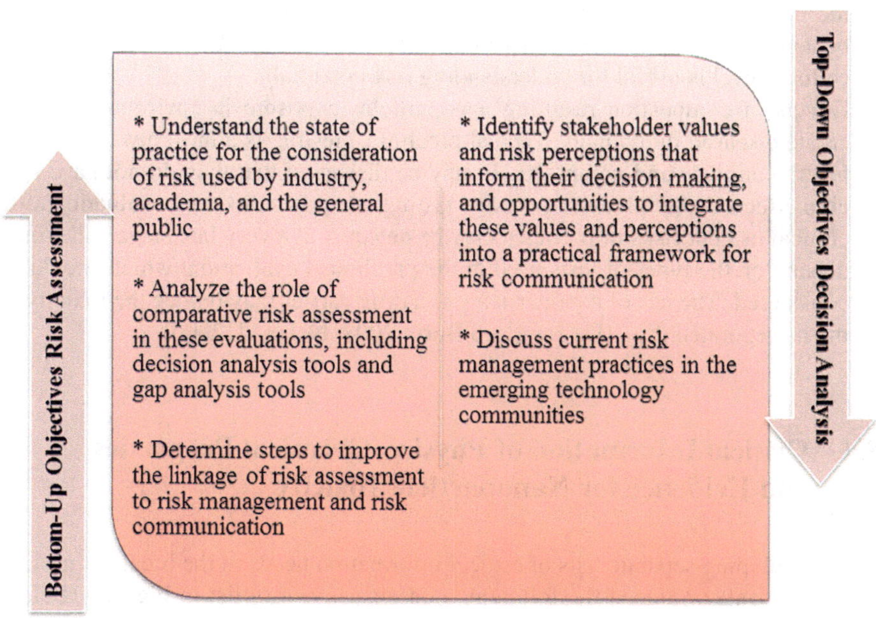

Fig. 8.3 2013 National Nanotechnology Initiative workshop objectives

Workshops displayed characterization information to modify best management performance to manage uncertain dangers. Uncertainty in the valuation process was often related to a lack of standard methods or test procedures, and the inability to predict statements of a technology in early stages of development. Finally, best management practices should be introduced to ensure worker safety (Fadel et al. 2015).

To use the potential of nanotechnology in all industries, full attention is required for safety and toxicological issues. However, for specific application, a toxicological evaluation is necessary. This is particularly true for the application of nanoparticles for drug delivery. In these applications particles are used in the human body and environment, and some of these new applications are an important enhancement of health care (Buxton et al. 2003). Toxicologists claimed that new science, methods and protocols are necessary, and this can make these evaluations cost effective, facilitating new product development (Nel et al. 2006). However, the requirement for this application is now emphasized by the following theories:

1. Nanomaterials are modified for their unique properties such as surface area in comparison to bulk materials. Since contact surface with the body tissue, unique properties are investigated from a toxicological viewpoint. Although current tests and procedures in drug and device evaluation may be appropriate to detect many dangerous reactions, it cannot be expected that these analyses will detect all potential risks. So additional analyses may be required that depend on the type of particles used.
2. Nanoparticles have different physico-chemical characteristics compared to micro-sized particles, which may result in changed body distribution, passage of the blood pathway and brain barrier. A basic understanding of the biological behavior of nanoparticles in distribution in vivo and body both at the organ and cellular level is useful for understanding risk potential.
3. Effects of combustion resulting nanoparticles exposure in environmental that create diseased individuals. Typical preclinical testing is almost always done in healthy animals and volunteers. It may be discussed that they do not have specific effects on health during routine testing and post marketing evaluation after clinical use but risks of particles may be detected at a very late stage. All would depend on the types of tests used in the preclinical evaluation, which should be considered. The use of nanoparticles as a drug carrier may reduce the toxicity of the incorporated drug (De Jong and Borm 2008; Nel et al. 2006).

8.4 Critical Information of Physico-Chemical Properties and Evidence for Nanoparticle Toxicity

After developing separate lists of critical information needs in the breakout groups, the participants produced the following consensus priority list in Fig. 8.4 (Balbus et al. 2007).

The main characteristic of nanoparticles is their size in the zone between individual atoms or molecules and the corresponding bulk materials. These physicochemical properties of the material can create the opportunity for increased uptake and interaction with biological tissues (Kreyling et al. 2010). This combination of effects can produce adverse biological effects in living cells, although the extraordinary properties of nanoparticles may require a novel investigative method to evaluate their hazard potential and particle toxicology (Buzea et al. 2007; Oberdörster et al. 2005b). Inhaled or instilled ambient nanoparticles can induce pulmonary inflammation, oxidative stress, distal organ involvement, fibrosis, cytotoxicity and mediator release from lung target cells (Donaldson and Tran 2002).

Tissue and cell culture analysis show the physiological response and role of oxidative stress in the making of inflammatory cytotoxic cellular responses. Clinical and experimental studies indicate that a small size, a large surface area and an ability to generate reactive oxygen species (ROS) play an important role in the ability of nanoparticles to induce lung injury (Nel 2005).

However, particle coating, surface treatments, surface excitation by ultraviolet radiation and particle aggregation can develop the effects of particle size. It is possible, therefore, that some nanoparticles may create their toxic effects as aggregates or through the release of toxic chemicals (Liu et al. 2014).

The increase in surface area causes an increase in the potential number of reactive groups, such as ROS, on the particle surface. So the change in the physicochemical and structural properties could be responsible for a number of material interactions that could lead to toxicological effects and increase in oxidation (Oberdörster et al. 2005a, b).

Also, these properties could confirm specific surface groups that could function as reactive sites. The extent of reactive sites and their importance intensely depend on the chemical composition of the material. Surface groups can make nanoparticle

Fig. 8.4 Critical information needs for nanoparticle toxicity

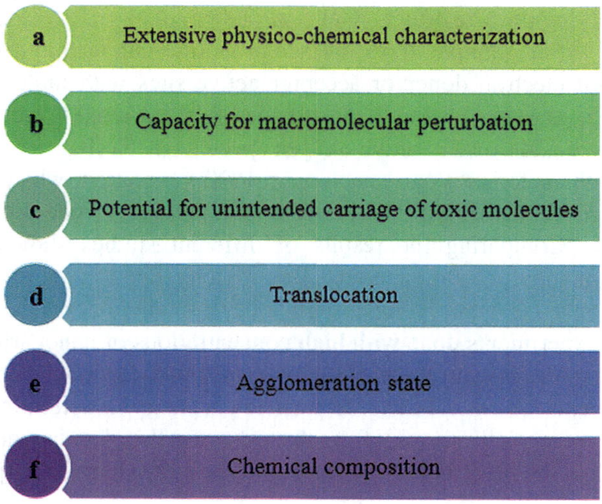

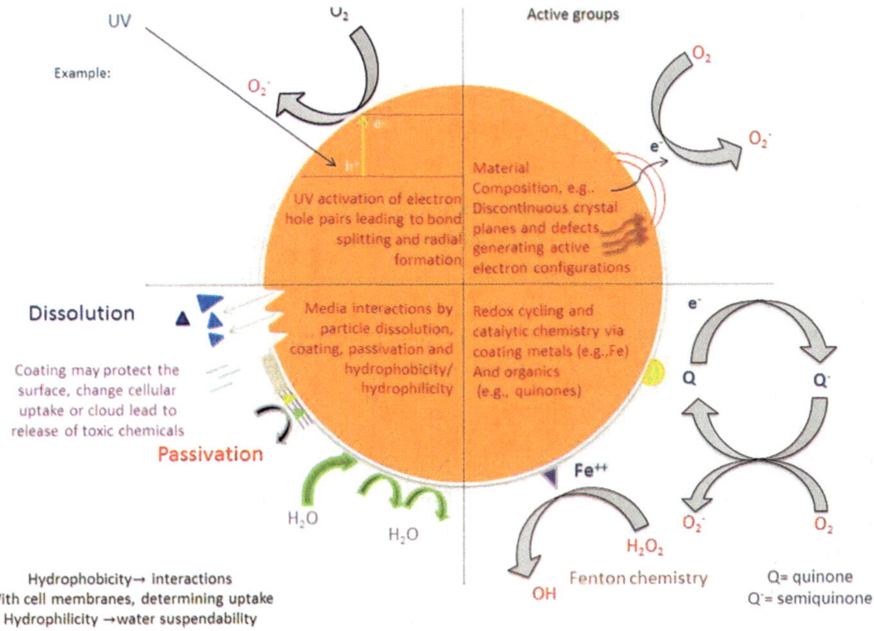

Fig. 8.5 Possible mechanisms by which nanomaterials interact with biological tissue

hydrophilic or hydrophobic, lipophilic or lipophobic, or catalytically active or passive, which are shown in Fig. 8.5 (Nel et al. 2006).

Toxicities of nanoparticles can be influenced by many other factors such as shape, aggregation, surface coating, surface charges, particle surface chemistry, biodegradability, concentration, and solubility may also affect the lectured specific physicochemical and transport properties with the possibility of negating or amplifying the size effects (Nel et al. 2006).

For example, those surface properties can lead to toxicity due to the interaction of electron donor or acceptor active sites with molecular oxygen (O_2). Electron capture can lead to the formation of the superoxide radical ($O_2^{\cdot-}$), which can generate additional reactive oxygen species (ROS) (Fig. 8.5). Thus, several nanoparticle characteristics can conclude in ROS generation, which is presently the developed model for nanoparticle toxicity (Table 8.1) (Shvedova et al. 2005).

Comparing the results of different studies showed that the aggregation of nanoparticles may reduce their toxicity, due to a more effective macrophage clearance for larger particles compared to smaller ones (Oberdörster et al. 2005b). Thus, experiments done with high concentrations of nanoparticles may not be as toxic as lower concentrations of the same nanoparticles (Buzea et al. 2007; Gurr et al. 2005). Another critical factor is particle chemistry in determining nanoparticle toxicity. It is especially relevant from the topic of cell molecular chemistry and oxidative stress. On the other hand, depending on their chemistry, nanoparticles can show different

Table 8.1 Nanoparticle' effects as the basis for pathophysiology and toxicity (Nel et al. 2006)

Experimental nanoparticle effects	Possible pathophysiological outcomes
ROS generation	• Protein • DNA and membrane injury • Oxidative stress
Oxidative stress	• Phase II enzyme induction • Inflammation • Mitochondrial perturbation
Mitochondrial perturbation	• Inner membrane damage • Permeability transition (PT) pore opening • Energy failure • Apoptosis • Apo-necrosis • Cytotoxicity
Inflammation	• Tissue infiltration with inflammatory cells • Fibrosis • Granulomas • Atherogenesis • Acute phase protein expression (e.g., C-reactive protein)
Protein denaturation, degradation	• Loss of enzyme activity • Auto-antigenicity
Nuclear uptake	• DNA damage • Nucleoprotein clumping • Autoantigens
Uptake in neuronal tissue	• Brain and peripheral nervous system injury
Altered cell cycle regulation	• Proliferation • Cell cycle arrest • Senescence
DNA damage	• Mutagenesis • Metaplasia • Carcinogenesis

cellular uptake and ability to catalyze the production of ROS. For example, rutile TiO_2 nanoparticles were found to induce oxidative DNA damage in the absence of light, but anatase TiO_2 nanoparticles of the same size did not (Wiesner and Bottero 2007). Moreover, hydroxyl radicals (·OH) associated with TiO_2 nanoparticles induced cytotoxicity and oxidative DNA damage (Reeves et al. 2008).

Also, surfactants can alter the physicochemical properties of nanoparticles and affect their cytotoxicity. Finally, besides the potential hazards of nanoparticles, their occurrence in the aqueous system are also important in determining their final toxicity to the public (Ahmad et al. 2005).

The challenge to relate the physicochemical properties of colloidal nanoparticles to their cytotoxicity is an important issue. So far, there is no agreement on the dose at which nanoparticles cause a biological response. Some of them measured the dose of toxicity by total weight, some others by the number of particles per volume. Some others found that the best way to find how toxic the particles are to cells was to calculate the dose based on the total surface area of the nanomaterial. Auffan et al. (2009) indicated that chemically stable metallic nanoparticles had no significant cellular toxicity, whereas nanoparticles able to be oxidized reduced or dissolved were cytotoxic and even genotoxic to cellular organisms. It seems that different parameters play major roles under different conditions on the toxicity evaluation of nanoparticles (Liu et al. 2014).

8.5 Hazard and Risk Evaluation for Nanoparticles

To determine the unique physico-chemical properties of nanoparticles, the nanotechnology community should try new ways to evaluate hazard and risk (Nel et al. 2006). These new approaches must also consider studies of chemical mixtures. The major database on the toxicity of nanoparticles has been created from inhalation and explosion toxicology, that they can be great reasons for research (Balbus et al. 2007). The toxicological profile of nanoparticles has developed during the past decade because of an increase in their consumption (Kreyling et al. 2004).

Toxicity of nanoparticles is an important discussion where adverse effects of environmental particulate air pollution comes from several sources. Particle toxicology proposes that more particle surface equals more toxicity and has adverse effects such as exacerbations of respiratory disease and deaths as well as hospitalizations and deaths from respiratory andcardiovascular disease (Brook et al. 2004). Other adverse effect of nanoparticles is pulmonary inflammation. Large surface areas which bind these adverse effects and the ability of nanoparticles together to cause inflammation can be seen as an important factor. Some nanoparticles may have the extra potential to affect cardiovascular disease directly (Mills et al. 2005). However, information of toxicity of nanoparticles in the body is limited, and all studies of nanoparticles have not shown clear information of toxicity from lung to the blood (Borm and Kreyling 2004; De Jong and Borm 2008).

One of the essentials in understanding nanoparticle toxicity is to understand the kinetics of inhaled ambient air nanoparticles. There are numerous mechanisms that are either based on the large surface area of particle core or on soluble components released by the nanoparticles whereby nanoparticles could lead to inflammatory effects. In addition, various chemicals, including those of biological origin like endotoxin—a component of the outer membrane of Gram-negative bacteria, may be adsorbed onto the nanoparticles and released in the body (Kreyling et al. 2004). In fact, a large surface area of organics and metals is characteristic of nanoparticles, so these have attracted significant toxicological attention in vitro. The aggregation of multiple chemical species, including biological compounds like endotoxin, limits the application of nanoparticles because of decreasing surface area (De Jong and Borm 2008).

As already mentioned above, nanoparticles exert some special properties that are very relevant in the further design of the development of engineered nanomaterials. Several effects are just quantitatively different from fine particles. Nanoparticles could also cause new types of effects, for example mitochondrial damage, uptake through olfactory epithelium, platelet aggregation, and cardiovascular effects where clearly new ways are a requirement for control over their toxicology. So advanced toxicological testing models should be used (De Jong and Borm 2008).

Risk characterization is defined as the final step to evaluation of risk procedure as an estimation of the incidence of the adverse effects to happen in a human or environment due to actual or predicted exposure to a substance (Hristozov and Malsch 2009). A basis of nano risk has been expanded by two organizations, namely Environmental Defense and the DuPont Company, for recognizing EHS risks. First, definition of health or environmental risks needs a realizing of both hazard and exposure based on reasonable attentions of the product lifecycle. In addition, the EHS framework could include hazard studies which provide a reasonable evaluation of the toxicity of the nanoparticles for human and environmental.

The nano risk framework contains six basic steps corresponding to stages of expanding. Briefly, these six steps are listed in Fig. 8.6.

For the user of this nano risk framework, step 2 contains three sections, a, b, and c. Section a is about identifying and characterizing the physical and chemical properties of the nanomaterial. It characterizes the hazard profile and identifies the nanomaterial's potential safety, health and environmental hazards in section b. Section c is planned to identify and characterize the potential for human or environmental exposures to the nanomaterial. For example, the results of base toxicity tests demonstrated that exposure to TiO_2 nanoparticles in the lungs of rats produced low

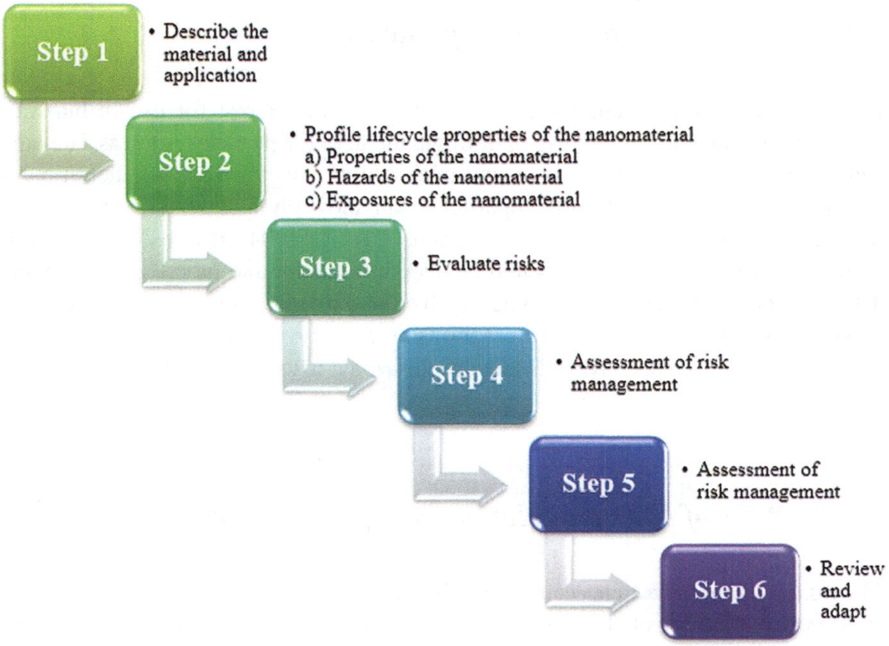

Fig. 8.6 Steps of the nano risk framework

inflammatory potential. These findings establish confidence for EHS attentions in expanding TiO_2 nanoparticles. Also the Nano Risk Framework approach shows reasonable assurances that the commercialization of this product has a very low health and environmental risk potential (Warheit et al. 2008).

Although the risks hazards of nano particles relevant to humans and the environment have been studied. A majority of the research has been done on animals which cannot be an accurate risk estimate for humans as there are fundamental differences between animals and humans against toxicity. Some of the risk of nanoparticles are as follows (Chalupa et al. 2004):

8.5.1 Risks of Inhaled Nanoparticles

The high deposition efficiency of inhaled nanoparticles in the pulmonary region increases in people with asthma or chronic obstructive lung disease. Inflammation of the lung is often seen as a response to the inhalation of nanoparticles as well (Chalupa et al. 2004). Warheit et al. (2007), by studying in vivo pulmonary toxicity in rats, showed that TiO_2 nanoparticles had low inflammatory potential and lung tissue toxicity. There was also evidence that nanoparticles might act as an adjuvant for allergic sensitization (Kreyling et al. 2004; Reijnders 2006).

8.5.2 Risks of Contacted Nanoparticles

Dermal penetration of nanoparticles is a material of interest for use in humans. Currently the most important use of ultrafine metal oxide particles, such as TiO_2 and ZnO in personal care products like sunscreens, highlights the dermal penetration of nanoparticles. TiO_2 and ZnO nanoparticles can show photocatalytic activity upon exposure to sunlight (Liu et al. 2014). Menzel et al. (2004) established in experiments that TiO_2 and ZnO may penetrate deep into the rat and rabbit skin. ZnO and TiO_2 nanoparticles may also become involved in damaging nucleic acids and other cell components by photocatalytic reactions on exposure to sunlight due to penetration into the granular layer. Nanoparticles may also cause allergic reactions (Reijnders 2006).

8.5.3 Risks of Nanoparticles in the Aquatic Environment

The toxicity of nanoparticles can be greatly affected by various factors in the aquatic environment. Bilberg et al. (2012) investigated critical toxicity of nanosilver on the body of a zebrafish in a 48-h static renewal study and compared their results with the toxicity of silver ions ($AgNO_3$). Results confirmed that silver nanoparticles were fatal compared to silver ions. Zhu et al. (2010) found that TiO_2 created minimal toxicity to daphnia within the traditional 48-h contact time, but caused high toxicity when the contact time was increased to 72 h. Their results demonstrated that contact duration may be an important factor in nanoparticle toxicity. A considerable amount of TiO_2 was also another effective factor that accumulated in daphnia, and these daphnia displayed difficulty in eliminating TiO_2 from their body.

8.6 Toxicity Pathways

In this section toxicity pathways will be discussed, which is relevant to environmental exposure to nanoparticles. Particle toxicology has been the subject of a number of reviews, considering the suitable approach for exposure, *in vitro* versus *in vivo* dose evaluations and its historical improvement (Driscoll et al. 2002). Five expressions have to be taken into account for the interpretation of inhaled particle properties: dose, deposition, dimension, durability and defense. First, the dose at a specific site of the body assesses the potential toxicity. This deposited dose is dependent on the concentration and size of the particle. In the respiratory system, the deposition probability of nanoparticles increases with decreasing particle size (Bailey and Roy 1994). If a particle is neither soluble nor degradable in the lung, it has a high durability and will accumulate. However, the lung has extensive defense systems, such as mucociliary clearance and macrophage clearance, to remove deposited nanoparticles. In addition, particle transport by macrophages from the alveolar region towards the larynx is slow in humans, even under normal conditions, thus eliminating only about a third of the deposited particles in the lung periphery. So the other two thirds, which accumulate in the lungs, are biodegradable and cleared by other mechanisms (Gehr and Heyder 2000). If particles are reactive or present sufficient dose, macrophages and epithelial cells can be damaged, leading to inflammation (Borm and Kreyling 2004). Nanoparticles can also translocate through the circulatory, lymphatic and nervous systems to many tissues and organs, including the brain (Buzea et al. 2007).

A main hazard study is to research human health and the so-called 'toxicokinetics' of nanoparticles related to exposure via the lung, skin and intestine (Fig. 8.7). Toxicokinetics investigates how a particle may get into the body, how it is circulated and dispersed within it, and how it may be metabolized and excreted. Understanding about toxicokinetics is important as it permits attention to the important target organs that may or may not be affected and how the body responds to nanoparticle exposure of metabolism and excretion (Khan 2014).

Studies have shown that patients with existing vascular disease have an increased risk of death. Figure 8.8 shows that effects can be classified into direct effects and indirect effects of nanoparticles on the heart and vessels. Among indirect effects, inflammation can be affected on target organs by mediators that are always available. A number of mechanisms have been assumed which are listed in Fig. 8.9.

The natural environment is made up of many complex ecosystems including atmospheric, terrestrial, fresh water, ground waters, soils and aquatic compartments.

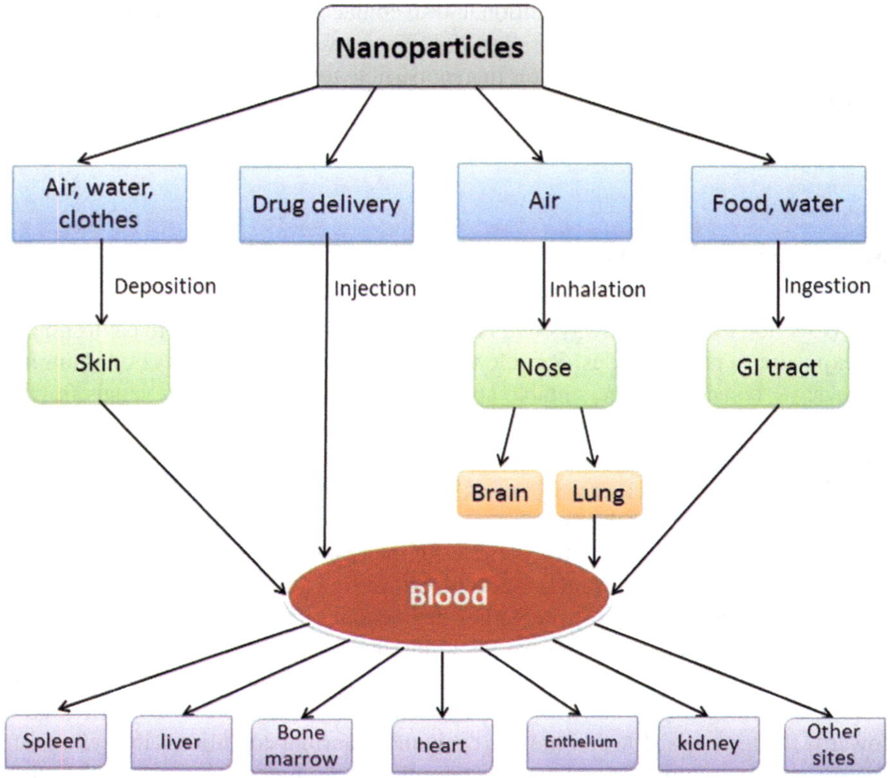

Fig. 8.7 Assumptive toxicokinetic pathway for nanoparticles

Unlike human exposure, the number of species potentially at risk from nanoparticles is extremely large. Hazards related to nanoparticle exposure can potentially act at individual or population level and might also impact on the structure and function of the ecosystem as a whole (Khan 2014; Leite et al. 2015).

8.7 Screening Hazards Test of Nanoparticle Applications

Since one of the new applications of nanoparticles is in novel drug delivery methods, nanoparticles are used to specific tissues and to increase drugs biological half-life. Also, a number of nanoparticles are used for imaging purposes for extravasation or tumor vascularization (De Jong and Borm 2008). A successful nanoparticle for drug conjugation and delivery must be able to have its pharmacokinetics developed at the organ and cellular level to promote increased delivery or efficiency and avoiding toxic effects of the nanoparticles. In fact, the lung as a barrier is bypassed in most applications and these nanoparticles should be considered to reach all target

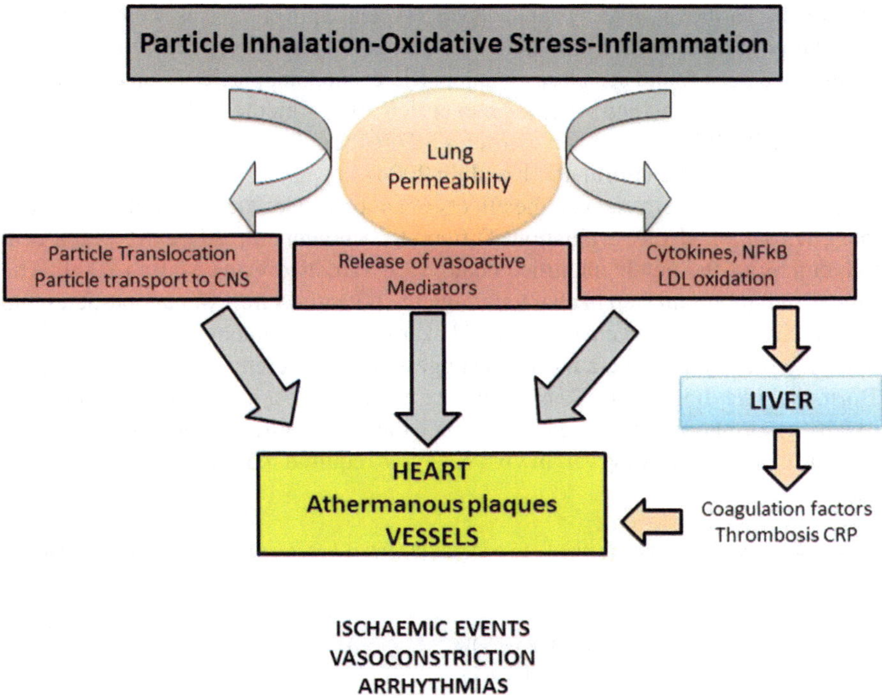

Fig. 8.8. Schematic of presentation of mechanisms and their sequence suggested through lung permeability

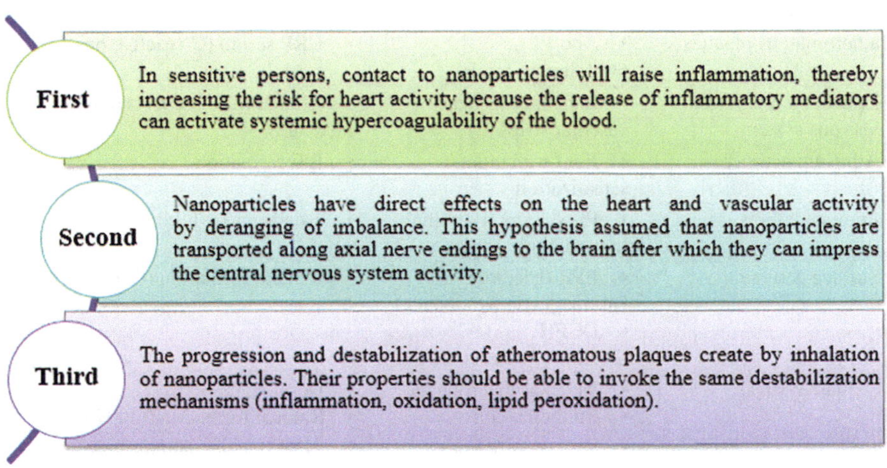

Fig. 8.9 Three mechanisms have been assumed for indirect effects of nanoparticles

cells of the cardiovascular system, hepatic system and kidney, as well as interact with the pool of immune competent cells. Several cells and/or receptors on these organs have been found sensitive to nanoparticle, and nanoparticles can disturb normal cell function in various ways. An overview of test methods that can be used to explore potential hazards of nanoparticles before their use as drug delivery tools can be found in Table 8.2 (Borm and Kreyling 2004).

Also, the ecotoxicity data on the effects of nanoparticles are essential for the suitable environmental risk evaluation. Different documents already exist which deal with emerging and newly identified health risks (Hristozov and Malsch 2009). The critical and important step in characterizing the potential safety of nanoparticles and the associated health and environmental hazards is development of a hazard outline. So a base set of hazard data has been suggested as a reference for characterization (Borm and Kreyling 2004). To characterize nanoparticles, its potential hazards and toxicity information on human health and the environment should be evaluated (Drobne et al. 2009). However, in vivo testing is required to rapidly identify hazard-

Table 8.2 Overview of test methods to explore potential hazards of nanoparticles (Borm and Kreyling 2004)

Hazard	Test system	Biomarker
Acute damage and responses after iv administration	• Red blood cells • Hemoglobin in supernatant	• Whole blood system • Release of inflammatory markers
Acute phase response	• Hepatocytes • Lung cells • Liver perfusion	• Fibrinogen • C-reactive protein, factor VII • Heat shock proteins
Increased permeability	• Endothelial or epithelial cell layers	• Permeability of H_2O or DTPA (probe to measure permeability)
Destabilization of atheromatous plaques	• Watanabe-rabbit • Apo-E (AplolipoproteinE)	• Plaque morphology • CRP serum (C-reactive protein)
Effects of autonomic nervous system	• Langendorff heart vascular strips	• Altered response to endogenous mediators
Adjuvant activity	• Rat Ovalbumin sensitization	• IgE formation
Immune effects	• PLN assay (Poplithal Lymph Node assay)	• T-cells and cytokine profiles
Surface activity	• EPR (Electron Paramagnetic Resonance) • DCFH (Dichlorofluorescine)	• Radical formation (OH, O_2, H_2O_2)
Oxidative stress	• Alveolar macrophage • Epithelial cell	• Isoprostane • Radical formation (O_2, H_2O_2)
Toxicity	• Various cell types • Primary and lines	• LDH (lactate dehydrogenase) release • MTT conversion dye-exclusion (dimethylthiazol-diphenyltetrazolium bromide)

ous substances with no available toxicity data. The TiO_2 nanoparticles has a number of applications such as in water treatment, food, sensors, additive pharmaceuticals and cosmetics (Masciangioli and Zhang 2003). It had been considered biologically inert prior to studies with nanoparticles which showed that TiO_2 nanoparticles activated an inflammatory response in laboratory test organisms (Drobne et al. 2009).

The extensive use of the nanoparticles brings the risk of chronic exposure to increasing levels of these compounds. Furthermore, their small size enables them to reach intracellular structures, such as under special circumstances the nucleus that are inaccessible for larger sized materials. Particle dissolution plays a critical role in hazard generation by metal and metal oxide nanoparticles and can have common effects on cell physiology with reactive oxygen species (ROS) (Soenen et al. 2015).

Another general observation is that toxicity is closely related to size and crystal structure, where nanoparticles are generally found to be more toxic than micrometer-sized materials (Beyerle et al. 2009; Soenen et al. 2015). In some cases, nanoparticles can be degraded totally by the biological processes in the environment, so their physical or chemical properties change. The mechanisms and potential for biodegradation is strongly dependent on the material properties that are still not fully understood. Most of the nanoparticles, such as ceramics and metal oxides, are not easily biodegradable (Hristozov and Malsch 2009).

The purpose of this section is to examine traditional toxicology tests. The term 'traditional' refers to those toxicology tests with well-established protocols that have been used in risk evaluation settings for many years. The term 'apical' refers to the fact that the end points evaluated in these tests may be the ultimate manifestation of a variety of different pathways of toxicity. Some participants suggested that the use of molecular technologies, including toxic genomics, may be more efficient at identifying novel mechanisms of toxicity. An additional, absorption-distribution-metabolism and excretion (ADME) method is developed. The general framework for ADME should be the same for nanoparticles, but the analytical challenges of detecting nanoparticles, as well as their by-products if they are transformed or metabolized, are significant. Some nanoparticles have special properties that make their real-time in vivo imaging straightforward. However, such imaging is not quantitative and must be confirmed by tissue analysis tests. Additionally, heavy reliance on imaging may not capture the dissolution or transformation of particles or their coatings. Nanoparticles that do not have intrinsic fluorescence or other properties that aid visualization present a greater challenge to accurately determine their ADME characteristics (Balbus et al. 2007).

8.8 Adverse Outcome Pathways (AOP)

Recently, a pathway-based view has been promoted for use in both human health and environmental toxicological risk evaluation. This approach aims to incorporate the information generated by non-traditional toxicity testing and high screening technologies based on small organism systems into the process of chemical hazard evaluation. The theory of adverse outcome pathways (AOPs) has been suggested as a framework to organize the existing knowledge on the toxicity mechanisms and outcomes across levels of biological organization for support of mechanism-based risk evaluation (Lee et al. 2015).

An AOP is an analytical, sequentially progressive pathway that links a molecular-initiating event to an adverse outcome. Recently, AOPs have been recognized as a potential informational tool by which the implications of molecular biomarkers in environmental risk assessments can be better understood (Reijnders 2006).

The AOP framework has also been suggested as a way to show how human-focused analyses can be useful in identifying key predicting effects in animal species. Effects of some of the nanoparticles could eventually lead to death, particularly in a natural environment. They are also involved in muscle contraction, memory, heart, and central neuronal activities. As mentioned earlier, ROS can lead to oxidative stress, which could result in cytotoxicity, pericardial edema, or apoptosis at the organ level. These effects could potentially lead to death of the organism (Garcia-Reyero et al. 2014). It is appropriate that research focuses on the life cycle of nanomaterials to reduce hazards. One of the ways is limiting exposure to manufactured nanoparticles resulting from production such as using precipitators for smaller particles and using magnetic filters to precipitate magnetic nanoparticles. Also in case of biodegradation, degradation products display very low toxicity (Reijnders 2006).

Toxicology of ultrafine particles has adverse health effects with the exposure to ambient. The potential advantages of nanomaterials are enormous, but they may possess potential hazards. Therefore, more attention on the risk assessment with nanomaterials is required. Nevertheless, knowledge about occurrence and toxicity of nanoparticles is still lacking. In this chapter, we have indicated what effects of nanoparticles have been found by toxicologists. We have focused on the important issue of hazard and risk evaluation for nanoparticles, which are critical conditions for studying health and the environmental impacts of nanomaterials.

References

Ahmad A, Senapati S, Khan MI, Kumar R, Sastry M. Extra-/intracellular biosynthesis of gold nanoparticles by an alkalotolerant fungus, Trichothecium sp. J Biomed Nanotechnol. 2005;1(1):47–53.

Assa F, Jafarizadeh-Malmiri H, Anarjan N, Berenjian A, Ghasemi Y. Applications of chitosan nanoparticles in active biodegradable and sustainable food packaging. In: Kale SA, Durai PRT, Prabakar K, editors. Renewable energy and sustainable development. Hauppauge: Nova Science Publishers; 2015.

Auffan M, Rose J, Orsiere T, De Meo M, Thill A, Zeyons O, Proux O, Masion A, Chaurand P, Spalla O. CeO2 nanoparticles induce DNA damage towards human dermal fibroblasts in vitro. Nanotoxicology. 2009;3(2):161–71.

Bailey M, Roy M. Annexe E. clearance of particles from the respiratory tract. Ann ICRP. 1994;24(1–3):301–413.

Balbus JM, Maynard AD, Colvin VL, Castranova V, Daston GP, Denison RA, Dreher KL, Goering PL, Goldberg AM, Kulinowski KM. Meeting report: hazard assessment for nanoparticles—report from an interdisciplinary workshop. Environ Health Perspect. 2007;115(11):1654.
Beyerle A, Schulz H, Kissel T, Stoeger T. Screening strategy to avoid toxicological hazards of inhaled nanoparticles for drug delivery: the use of a-quartz and nano zinc oxide particles as benchmark. J Phys Conf. Ser. 2009;151:012034: IOP Publishing
Bilberg K, Hovgaard MB, Besenbacher F, Baatrup E. In vivo toxicity of silver nanoparticles and silver ions in zebrafish (Danio rerio). J Toxicol. 2012;2012:293784.
Borm PJ, Kreyling W. Toxicological hazards of inhaled nanoparticles—potential implications for drug delivery. J Nanosci Nanotechnol. 2004;4(5):521–31.
Borm PJ, Robbins D, Haubold S, Kuhlbusch T, Fissan H, Donaldson K, Schins R, Stone V, Kreyling W, Lademann J. The potential risks of nanomaterials: a review carried out for ECETOC. Part Fibre Toxicol. 2006;3(1):11.
Brook RD, Franklin B, Cascio W, Hong Y, Howard G, Lipsett M, Luepker R, Mittleman M, Samet J, Smith SC. Air pollution and cardiovascular disease. Circulation. 2004;109(21):2655–71.
Buxton DB, Lee SC, Wickline SA, Ferrari M. Recommendations of the National Heart, Lung, and Blood Institute Nanotechnology Working Group. Circulation. 2003;108(22):2737–42.
Buzea C, Pacheco II, Robbie K. Nanomaterials and nanoparticles: sources and toxicity. Biointerphases. 2007;2(4):MR17–71.
Chalupa DC, Morrow PE, Oberdörster G, Utell MJ, Frampton MW. Ultrafine particle deposition in subjects with asthma. Environ Health Perspect. 2004;112(8):879.
De Jong WH, Borm PJ. Drug delivery and nanoparticles: applications and hazards. Int J Nanomedicine. 2008;3(2):133.
Donaldson K, Tran CL. Inflammation caused by particles and fibers. Inhal Toxicol. 2002;14(1):5–27.
Driscoll KE, Carter JM, Borm PJ. Antioxidant defense mechanisms and the toxicity of fibrous and nonfibrous particles. Inhal Toxicol. 2002;14(1):101–18.
Drobne D, Jemec A, Tkalec ŽP. In vivo screening to determine hazards of nanoparticles: nanosized TiO_2. Environ Pollut. 2009;157(4):1157–64.
Fadel TR, Steevens JA, Thomas TA, Linkov I. The challenges of nanotechnology risk management. Nano Today. 2015;10(1):6–10.
Forbe T, García M, Gonzalez E. Potential risks of nanoparticles. Food Sci Technol (Campinas). 2011;31(4):835–42.
Foss Hansen S, Larsen BH, Olsen SI, Baun A. Categorization framework to aid hazard identification of nanomaterials. Nanotoxicology. 2007;1(3):243–50.
Garcia-Reyero N, Kennedy AJ, Escalon BL, Habib T, Laird JG, Rawat A, Wiseman S, Hecker M, Denslow N, Steevens JA. Differential effects and potential adverse outcomes of ionic silver and silver nanoparticles in vivo and in vitro. Environ Sci Technol. 2014;48(8):4546–55.
Gehr P, Heyder J. Particle-lung interactions. Boca Raton: CRC Press; 2000.
Guadagnini R, Moreau K, Hussain S, Marano F, Boland S. Toxicity evaluation of engineered nanoparticles for medical applications using pulmonary epithelial cells. Nanotoxicology. 2015;9(suppl 1):25–32.
Gurr J-R, Wang AS, Chen C-H, Jan K-Y. Ultrafine titanium dioxide particles in the absence of photoactivation can induce oxidative damage to human bronchial epithelial cells. Toxicology. 2005;213(1):66–73.
Howard J. Current intelligence bulletin 65: occupational exposure to carbon nanotubes and nanofibers. DHHS (NIOSH) Publication. 2013;2013:145.
Hristozov D, Malsch I. Hazards and risks of engineered nanoparticles for the environment and human health. Sustainability. 2009;1(4):1161–94.
Kavlock R, Dix D. Computational toxicology as implemented by the US EPA: providing high throughput decision support tools for screening and assessing chemical exposure, hazard and risk. J Toxicol Environ Health B Crit Rev. 2010;13(2–4):197–217.
Khan FH. Chemical hazards of nanoparticles to human and environment (a review). Orient J Chem. 2014;29(4):1399–408.

Kreyling WG, Semmler M, Möller W. Dosimetry and toxicology of ultrafine particles. J Aerosol Med. 2004;17(2):140–52.

Kreyling WG, Semmler-Behnke M, Chaudhry Q. A complementary definition of nanomaterial. Nano Today. 2010;5(3):165–8.

Lee JW, Won E-J, Raisuddin S, Lee J-S. Significance of adverse outcome pathways in biomarker-based environmental risk assessment in aquatic organisms. J Environ Sci. 2015;35:115–27.

Leite PEC, Pereira MR, Granjeiro JM. Hazard effects of nanoparticles in central nervous system: searching for biocompatible nanomaterials for drug delivery. Toxicol In Vitro. 2015;29(7):1653–60.

Liu Y, Tourbin M, Lachaize S, Guiraud P. Nanoparticles in wastewaters: hazards, fate and remediation. Powder Technol. 2014;255:149–56.

Masciangioli T, Zhang W-X. Peer reviewed: environmental technologies at the nanoscale. Environ Sci Technol. 2003;37:102A–8A. ACS Publications

Maynard AD, Aitken RJ, Butz T, Colvin V, Donaldson K, Oberdörster G, Philbert MA, Ryan J, Seaton A, Stone V. Safe handling of nanotechnology. Nature. 2006;444(7117):267–9.

Menzel F, Reinert T, Vogt J, Butz T. Investigations of percutaneous uptake of ultrafine TiO2 particles at the high energy ion nanoprobe LIPSION. Nucl Instrum Methods Phys Res B. 2004;219:82–6.

Mills NL, Törnqvist H, Robinson SD, Gonzalez M, Darnley K, MacNee W, Boon NA, Donaldson K, Blomberg A, Sandstrom T. Diesel exhaust inhalation causes vascular dysfunction and impaired endogenous fibrinolysis. Circulation. 2005;112(25):3930–6.

Nel A. Air pollution-related illness: effects of particles. Science. 2005;308(5723):804–6.

Nel A, Xia T, Mädler L, Li N. Toxic potential of materials at the nanolevel. Science. 2006;311(5761):622–7.

Oberdörster G. Safety assessment for nanotechnology and nanomedicine: concepts of nanotoxicology. J Intern Med. 2010;267(1):89–105.

Oberdörster G, Maynard A, Donaldson K, Castranova V, Fitzpatrick J, Ausman K, Carter J, Karn B, Kreyling W, Lai D. Principles for characterizing the potential human health effects from exposure to nanomaterials: elements of a screening strategy. Part Fibre Toxicol. 2005a;2(1):8.

Oberdörster G, Oberdörster E, Oberdörster J. Nanotoxicology: an emerging discipline evolving from studies of ultrafine particles. Environ Health Perspect. 2005b;113(7):823.

Reeves JF, Davies SJ, Dodd NJ, Jha AN. Hydroxyl radicals (OH) are associated with titanium dioxide (TiO2) nanoparticle-induced cytotoxicity and oxidative DNA damage in fish cells. Mutat Res. 2008;640(1):113–22.

Reijnders L. Cleaner nanotechnology and hazard reduction of manufactured nanoparticles. J Clean Prod. 2006;14(2):124–33.

Roco MC. National nanotechnology initiative-past, present, future. In: Goddard WA, et al., editors. Handbook on nanoscience, engineering and technology. Boca Raton and London: CRC, Taylor and Francis; 2007. 3.1–3.

Royal Society and Royal Academy of Engineering. Nanoscience and nanotechnologies: opportunities and uncertainties. London: Royal Society and Royal Academy of Engineering; 2004.

Salata OV. Applications of nanoparticles in biology and medici. J Nanobiotechnology. 2004;2(1):3.

Schmidt CW. Nanotechnology-related environment, health, and safety research: examining the national strategy. Environ Health Perspect. 2009;117(4):A158.

Shvedova AA, Kisin ER, Mercer R, Murray AR, Johnson VJ, Potapovich AI, Tyurina YY, Gorelik O, Arepalli S, Schwegler-Berry D. Unusual inflammatory and fibrogenic pulmonary responses to single-walled carbon nanotubes in mice. Am J Physiol Lung Cell Mol Physiol. 2005;289(5):L698–708.

Soenen SJ, Parak WJ, Rejman J, Manshian B. (Intra) cellular stability of inorganic nanoparticles: effects on cytotoxicity, particle functionality, and biomedical applications. Chem Rev. 2015;115(5):2109–35.

Vauthier C, Dubernet C, Fattal E, Pinto-Alphandary H, Couvreur P. Poly (alkylcyanoacrylates) as biodegradable materials for biomedical applications. Adv Drug Deliv Rev. 2003;55(4):519–48.

Warheit DB. Nanoparticles: health impacts? Mater Today. 2004;7(2):32–5.
Warheit DB, Hoke RA, Finlay C, Donner EM, Reed KL, Sayes CM. Development of a base set of toxicity tests using ultrafine TiO_2 particles as a component of nanoparticle risk management. Toxicol Lett. 2007;171(3):99–110.
Warheit DB, Sayes CM, Reed KL, Swain KA. Health effects related to nanoparticle exposures: environmental, health and safety considerations for assessing hazards and risks. Pharmacol Ther. 2008;120(1):35–42.
Wiesner M, Bottero J-Y. Environmental nanotechnology. New York: McGraw-Hill Professional Publishing; 2007.
Zhu X, Chang Y, Chen Y. Toxicity and bioaccumulation of TiO_2 nanoparticle aggregates in Daphnia magna. Chemosphere. 2010;78(3):209–15.

Chapter 9
Commercialization Consideration

9.1 Introduction

Nanotechnology is assumed to have great potential in the future market. The most important feature of nanotechnology is its novelty on the path to commercialization. Inventions often attract attention in industrial and scientific communities because of their ingenuity, but a product must also be useful and practical. This chapter aims to review commercialization consideration of nanotechnology. This chapter also looks at the purpose and process of commercialization of nanotechnology for various applications.

Ability of any technology to design and produce useful and marketable products is its final purpose using available practical knowledge and skills. Nanotechnology is one of these technologies that has different applications in all fields (Lewin 2014). The term 'commercialization' is defined as changing or transferring 'technology' into a profit-making position. By technology, techniques, equipment and systems researchers and patents can be commercialized with appropriative processes. Commercialization of research and development (R&D) may also refer to the economic impact of nanotechnology (Fiedler and Welpe 2010; Lewin 2014).

Researchers up until now have studied the economic potential and market applications of nanotechnology in a variety of different fields such as electrical and chemical engineering, food, medical instruments and biotechnology (Fiedler and Welpe 2010).

Nanotechnology is estimated to play an important role with market potential in numerous applications from healthcare to energy systems (Pomykalski and Bakalarczyk 2011).

Governments, and some important organizations, such as international standards governing agencies, environmental, non-governmental organization (NGOs), and laypeople regard nanotechnology as a problem. They are concerned about the potential negative long-term consequences of nanomaterials on human health and

H. Jafarizadeh-Malmiri et al., *Nanobiotechnology in Food: Concepts, Applications and Perspectives*, https://doi.org/10.1007/978-3-030-05846-3_9

the environment. The commercialization of products and processes utilizing nanotechnology is now on the political schedule of national governments and international organizations (Lewin 2014; Youtie et al. 2008).

It is known that innovation, when commercialized, is an important contributor to the growth of the economy. The available literature on technology entrepreneurship and commercialization strategies is huge (Walsh 2012).

Viewpoints of nanotechnology management studies have also changed. Recently, many studies are concentrating on the factors, either positive or negative, that affect the commercialization of nanotechnologies on human and environmental life. Previously, most studies focused on comparing nano and bio technologies, on characterizing nanotechnology as a general objective technology, or on analyzing the competitiveness of various nations in nanotechnology (Lee et al. 2013; Roco and Bainbridge 2002).

This chapter is about nanotechnology commercialization, market entering strategies and an overview of some of the early commercial efforts using nanotechnology in the life sciences.

9.2 The Purpose of Commercialization

Years of research and development created new products, new knowledge and therefore intellectual property. Industry's main interest and purpose of commercialization is creating profit so possession of intellectual property is an essential beginning of the commercialization process. This is not the key purpose or first interest of academia since it derives most of its funding support from government to transfer patterns to high income (Khosravi and Sadeghi 2014; Tolfree and Jackson 2007).

The key obstacle to developing any commercialization approach is the lack of an appropriate infrastructure that supports research and product development (Tolfree and Jackson 2007).

Nikulainen and Palmberg studied the relationships between the purposes of researchers, university–industry interactions, and nanotechnology transfer challenges. They demonstrate that the most essential modes of industry–university interactions in the field of nanotechnology take place in public R&D programs (Nikulainen 2010). Figure 9.1 comprises four categories and ten key factors as drivers or hurdles for nanotechnology commercialization (Lee et al. 2013).

9.3 The Process of Commercialization

To develop commercialization in nanotechnology, there is a need to define steps in more detail with regard to the multiple sub-fields within the nanotechnology domain (Porter et al. 2008). Governments, universities (through their technology

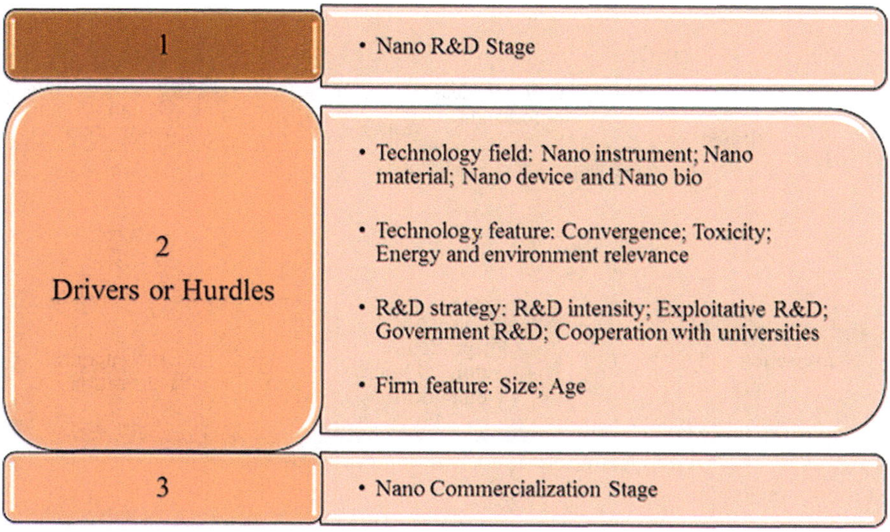

Fig. 9.1 Research framework and categories of nanotechnology

transfer offices) and the industry (companies and start-ups) are commercial agents that want to commercialize nanotechnology (Lewin 2014). The important steps to take an idea for a new product to the market are shown in Fig. 9.2. There is not a path for commercialization without determination, knowledge and experience. Experience shows that a suitable organization is needed, otherwise it is impossible to take the steps on the path to commercialization (Oriakhi 2004; Tolfree and Jackson 2007). There is a large gap between an idea and commercialization. Most ideas do not cross this gap. For commercialization of ideas, research, focused determination and specialized technology knowledge are essential (Tolfree and Jackson 2007).

Also, Fig. 9.3 shows the flow steps of nanotechnology commercialization (Lewin 2014).

Despite all the issues and concerns about nanotechnology, investments in the market of nano-tech-based products are predicted to increase. Proposed steps of a nanotech product from conception to a successful commercial venture are shown in Fig. 9.4 (Kaur et al. 2014).

Industry requirements agreement on standards can grow before any new developed methodology. Developed standards are a continuing issue for the Micro Electrical Mechanical Systems/Micro Systems Technology/Nano Electrical Mechanical Systems (MEMS/MST/NEMS) industries. Also there are various standard development organizations worldwide such as Semiconductor Equipment, Materials International (SEMI) and the International Organization for Standardization (ISO) (Shapira and Wang 2009; Tolfree and Jackson 2007).

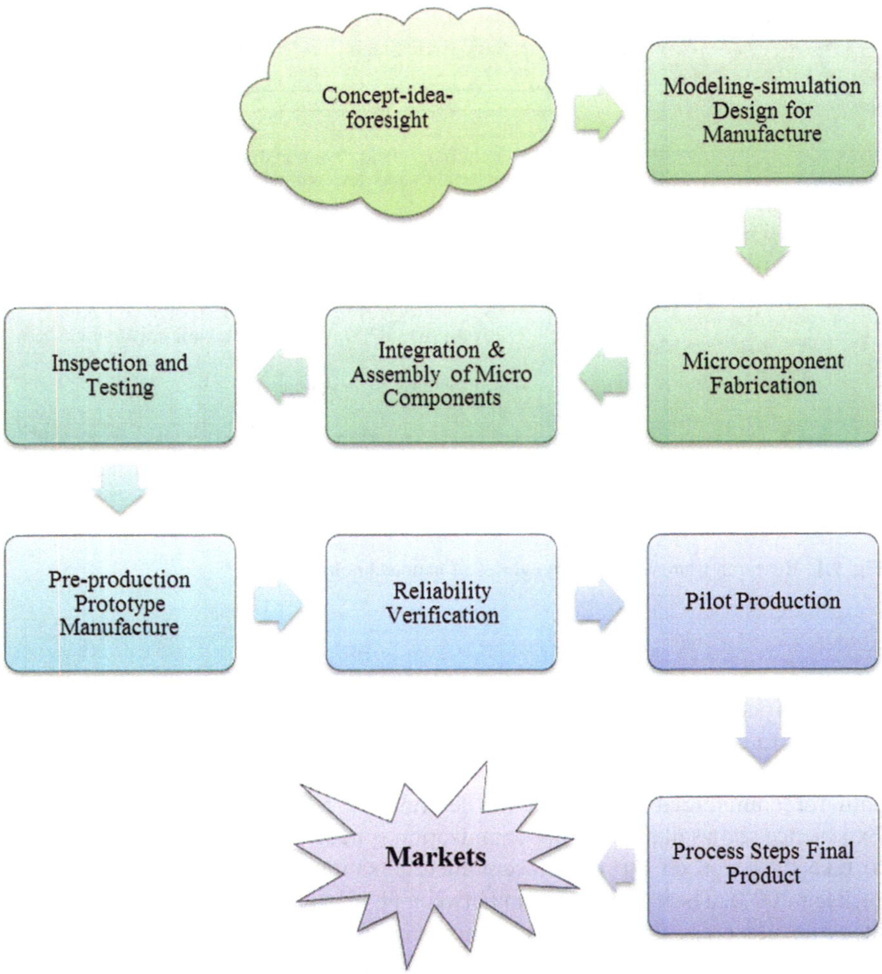

Fig. 9.2 The important steps to take an idea to the market

9.4 Commercialization Strategies for Research, Patent and Nanotechnology Start-Ups

Despite risks, and the current limits of public engagement in discussions about nanotechnology, scientists and companies across all sectors of the agri-food system study and research nanotech materials and products to commercialize (Scrinis and Lyons 2007). It is necessary to reach firms that increase amounts of basic research. Patents are a fundamental base of what is commercialized in the bioscience and nanoscience sectors, although there are enormous parts of patents that are not commercialized. It is essential to know how licenses are evaluated and commercialized (Gómez Uranga et al. 2007).

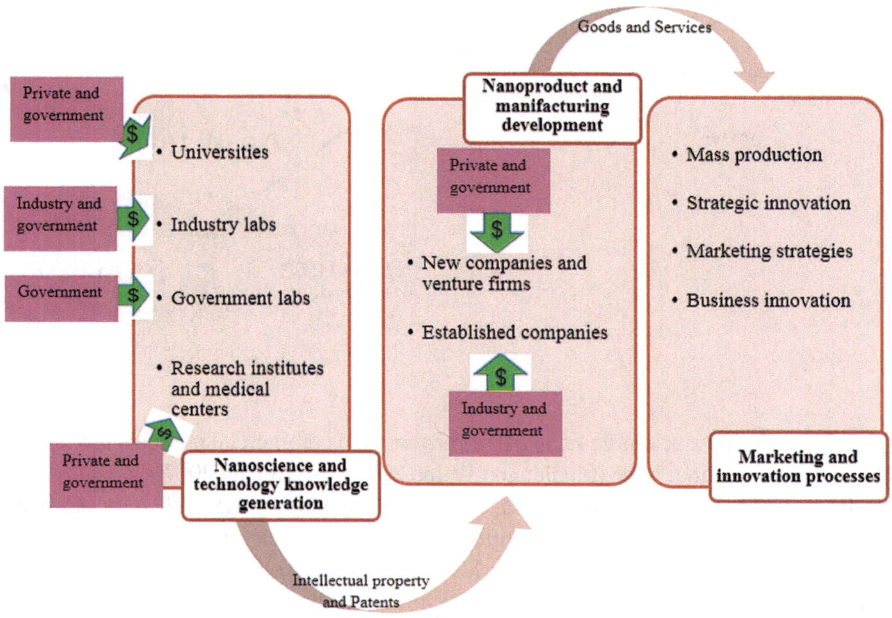

Fig. 9.3 Flow steps of nanotechnology commercialization

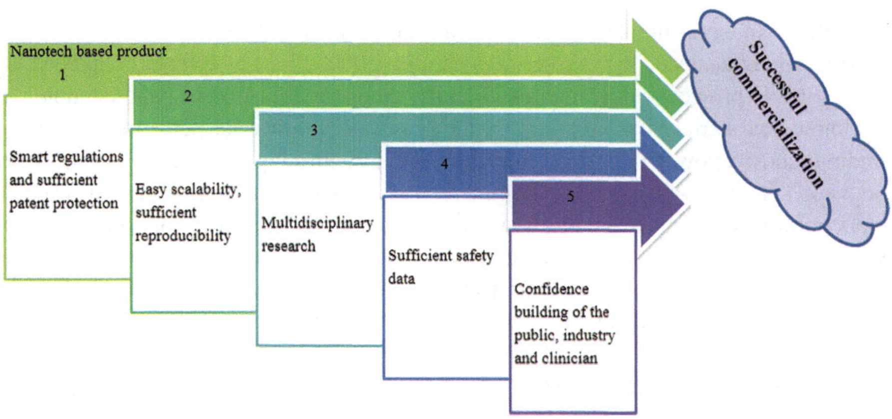

Fig. 9.4 Proposed steps of a nanotech product from conception to a successful commercial venture

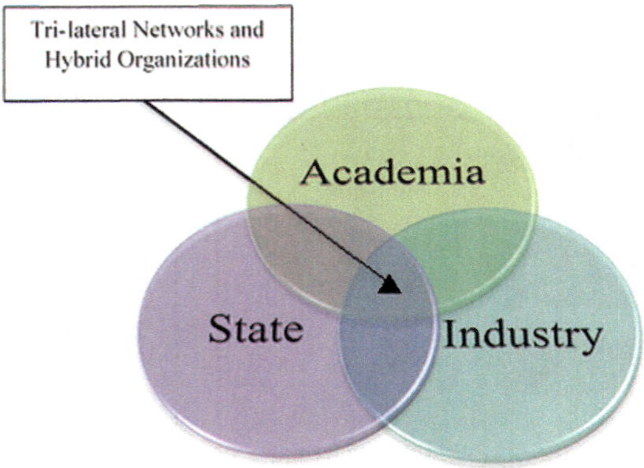

Fig. 9.5 The Triple Helix Model (THM) of university, industry and government

Understanding how scientific knowledge converts into commercial products in the market is a main research issue (Fiedler and Welpe 2010; Meyyappan 2007).

Because of enormous commercial potential, nanotechnology is developing from scientific idea to market. For example, market size of nanotechnology was estimated at over US$1 trillion total worldwide in 10–15 years. Also, about 2.5% of scientific articles and 0.7% of patents are related to nanotechnology topics, which can be seen as a suggestion that the commercial potential of nanotechnology will have at least the same degree as biotechnology (Von Raesfeld et al. 2012).

Figure 9.5 shows a visualization of the Triple Helix Model (THM) of university, industry and government. This idea has been introduced by Leydesdorff and Etzkowitz. The use of the THM provides an approach to show the interrelationships among innovation, the commercialization processes and the parties involved so it is suitable as an experimental model (Tolfree and Jackson 2007).

Raesfeld et al. tested hypotheses on a dataset (2000–2008) of 169 public nanotechnology R&D projects, which are aimed at the commercialization of nanotechnology. Technological variety has an effect on the projects' commercial performance. Their findings show a powerful positive impact on value chain supplementary of partners on both application development and commercial performance of the projects. The framework introduced in their study allows an evaluation of the effects of technological variety and value chain supplementary on application development and the commercial performance of public R&D projects (Von Raesfeld et al. 2012). From 1990 to 2006, U.S. universities and research organizations recorded over 90,000 nanotechnology publications and over 11,000 nanotechnology patents, whereas industry produced about 10,000 nanotechnology publications and some 50,000 nanotechnology patents that are shown in Fig. 9.6. In the U.S., industry is the major player in invention through patenting (Rasmussen 2007; Shapira and Wang 2009).

Study on the technology entrepreneurship field includes the overlap of technological innovation and entrepreneurship. There are a number of strategies in this field, one of them is strategies required for technology companies to provide a return on their innovation (commercialization). The type of innovation and the related commercial risk, such as market risk, product risk and cost risk, related with introducing that innovation to the marketplace, influence the choice of commercialization strategy (Walsh 2012). Table 9.1 shows some of the commercial strategies of nanotechnology start-ups.

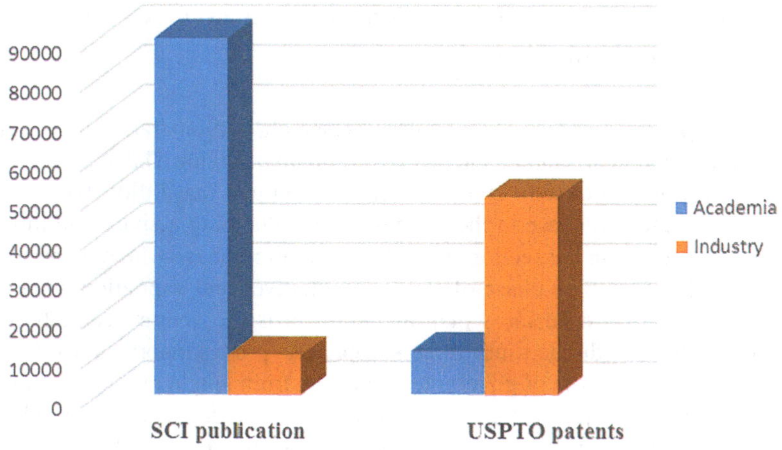

Fig. 9.6 Nanotechnology R&D indicated by publications and patents 1990–2006. *SCI* Science Citation Index publication, *USPTO* United States Patent and Trademark Office

Table 9.1 Commercial strategies of nanotechnology start-ups

No.	Commercial strategies	Commercial barriers
1	• Assistance upwards and downwards the value chain • Relationship with the public sector • Superb technology knowledge	• Market accept
2	• Build core competence • Technology knowledge • Networking • International collaboration	• Market accept • Market restrictions
3	• Strategic partner • Relationship with the public sector • Build core competence • Find suppliers	• Market access • Market accept of the technology
4	• Technology ownership • Collaboration with the public sector • Pilot customer • Technology follower	• Cost often no match between investors and nanotechnology start-ups
5	• Technology ownership • Clear definition of the technology • Proprietary technology • Unique technology • First mover	• Time to market
6	• Clearly definable, quantitative, value proposition • Unique technology • Refine the technology • Fine-tune the innovation	• Time to market • Established • Companies may have legacy issues

9.5 Market Entering Strategies and Potential Growth for Nano Products in Future

Table 9.2 shows three phases of technology: research, demonstration and commercialization for technical, market and business domain behavior. The domains change in parallel responding to each other. The assessment and feasibility study is in the section of the research phase in the three domains. Planning activity and introduction of the product is in the section of the demonstration phase which is the stage of prototyping. Market is the phase of strategic initiatives and validation. One of the important stages of the business plan and start-up is the venture capitalists. The commercial phase is divided into three sub-phases, pre-, support- and fully commercial. However, the role of government changes from single procurer to a noninterfering agency as the product gains self-sufficiency. During the commercialization phase the production can reach full scale, market gets integrated and business grows and develops (Balachandra et al. 2010; Frewer et al. 2014).

A wide range of nanomaterials is used for consumer products and various new products that have been launched in the market. Nanomaterials have been applied in various industries, for example the food and drug industry that are explained below (Bumbudsanpharoke and Ko 2015).

9.5.1 Medical Market

Successful commercialization of nanotechnology in medical science has already happened, which introduced new products, advanced processes and procedures that could have not improved significantly without the use of specifically designed nanotechnologies (Hobson 2009).

Metallic oxide sunscreen components, such as zinc oxide and titanium dioxide, were the first commercialization of nanomaterials in medical science. Also, nanoparticulate silver was very successfully commercialized for its antimicrobial property many years ago. The universal commercialization for the detection, development and marketing for nanomedical was shown in Fig. 9.7 (Bumbudsanpharoke and Ko 2015; Hobson 2009).

Table 9.2 Phase of technology-research and commercialization

Phase	Sub-phase/stage	Principal activity	Technical	Market
Research	Basic research Applied research	Feasibility study	Technology feasibility	Market study Economic study
Demonstration	Design and development Engineering	Planning Introduction	Engineering prototype	Strategic marketing
Commercial	Pre commercial Support commercial Fully commercial	Self-sufficiency	Small scale production	Market diversification

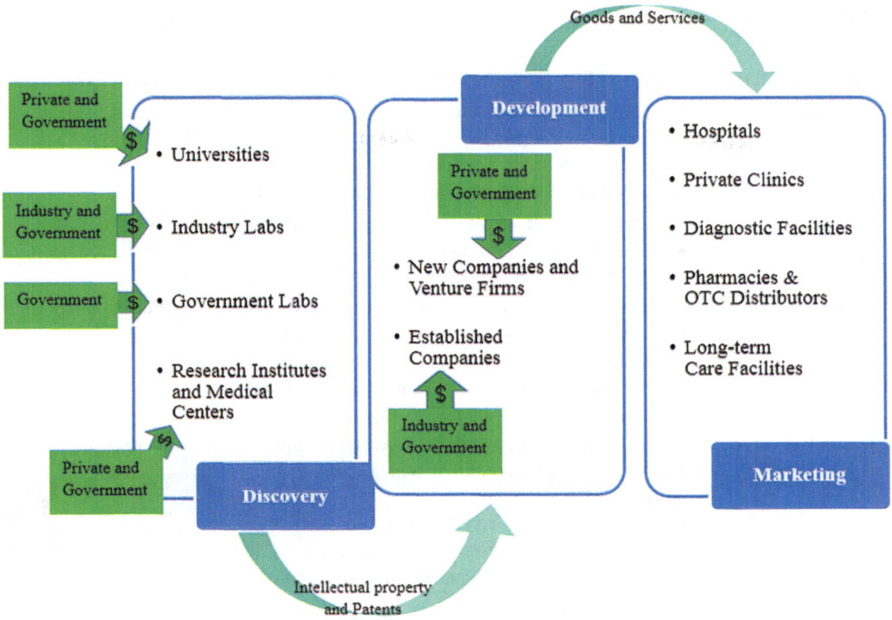

Fig. 9.7 Commercialization discovery, development for medical nanotechnology

There are only a few concerns about efficacy and safety in commercializing medical nanotechnology. So studies are needed to obtain some unique data as well as demonstration of the ability to safely manufacture. Therefore, various aspects critical to the successful commercialization of nanotechnologies in medicine have to be considered (Hobson 2009).

The National Science Foundation reported that the worldwide market size for nanoproducts was $1 trillion in 2015. With respect to pharmaceuticals alone, this report shows that the impact of nanotechnology was about US$180 billion per year around 2010–2015 (Bumbudsanpharoke and Ko 2015; Hobson 2009).

Thus, based on these data, the market potential for new pharmaceutical nanotechnologies is expected to become increasingly significant. The world pharmaceutical market is estimated to reach US$1.3 trillion by 2020 (Hobson 2009).

9.5.2 *Food Market*

It is estimated that up to 400 companies around the world use nanotechnology in the food industry. According to a Persistence Market Research (PMR) report, the global nano enabled packaging market for food was estimated to be US$6.5 billion in 2013 and is expected to reach about US$15.0 billion in 2020 at a compound annual growth rate of 12.7%. On the other hand, the European Institute guesses that the use

Table 9.3 Examples of nano-food packaging in the market (Bumbudsanpharoke and Ko 2015)

Nanomaterial type	Improved functionality from product claim	Application or product image
Nanoclay	• Barrier nylon resin provides an excellent barrier • Barrier nylon resin is well suited to the co-injection process because its recommended processing temperature is similar to that of PET	• Drink juice bottles
Nanoclay	• Imperm R can with a more cost effective material that allows for easier processing and maintaining barrier properties	• Bottles from Miller Brewing (U.S.A.)
Nanosilver (particles size 25 nm)	• Keep foods fresher 3 or even 4 times longer for fruits, vegetables, herbs, breads, cheeses, soups, sauces, and meats • Growth of bacteria containing nanosilver was reduced by over 98% in comparing	• Pocket of food
Nanosilver (size 20 to 70 nm)	• FreshBox is a newly developed nanosilver antimicrobial food container which is made by unique nanotechnology • FreshBox shows excellent antimicrobial properties against various bacteria and fungus due to the effect of finely dispersed nanosilver particles	
A silver-base zeolite antimicrobial agent	• Antimicrobial (bacteria, enzyme, and molds) • To kill pathogenic organisms, reducing their number to an extent that is not harmful, and making them harmless	• Plastic films for food packaging

of nanomaterials in the food packaging market is expected to reach US$20 billion by 2020. However, the entire nanotech industry was estimated to reach $1 trillion in 2015 (Bumbudsanpharoke and Ko 2015). Table 9.3 shows several examples of the available products in the market.

9.6 Factors Influencing Nanotechnology Commercialization

There is a copious amount of information about the factors that affect the successful commercialization of traditional technologies or new technologies which can then be used in various industries. Lately, analyzing the factors for success in commercializing technologies, especially information and communications technology, biotechnology, energy technology are important (Lee et al. 2013).

The following are five critical success factors for nanotechnology commercialization from a review of the literature and combined with experience gained:

1. Product orientation (and not technology admiration);
2. Continuous market interaction and selection of their practical application;
3. Application of product development methodology;
4. Attraction and retention of commercialization partners;
5. Mitigation of nanotechnology-specific technology risks (Tsuzuki 2013).

9.7 Economic Impact of Nanotechnology Commercialization

The economic impact of nanotechnology to medical applications has not been well studied. The most promising forecasts of a positive economic impact for nanotechnology predict the complete integration of nanotechnology into the present economy due to a high situation of preparedness, fundamental investments by numerous different businesses, and effective strategic design (Hobson 2009). Therefore, to have maximum positive economic as well as social impact, a realization of the strategic economic value on the healthcare budgets of nations and the role of nanotechnology applications would be required for the commercialization of nanotechnology in medicine (Zucker and Darby 2005).

9.8 Challenges and Barriers to Commercialization of Nanotechnology

A study by the US Department of Commerce was conducted with an in-depth analysis of the barriers to commercializing nanotechnology that listed 22 barriers, including:

1. Adverse public perception;
2. Expensive infrastructure for producing;
3. Food and drug administration and patent offices with inadequate qualified staff to assess nanotechnology products (Lee et al. 2013; Rasmussen 2007).

Commercialization of technology has various motivations. Governments and environmental organizations pay attention to the hopeful applications of the technology for solving environmental problems such as soil degradation and water scarcity. Public concerns about the environmental impact of nanotechnology on human health and nature need to be resolved before commercialization. There is also a lot of R&D on the knowledge gap with regard to releasing engineered nanomaterials in the nature. One of the major barriers to commercialization is the communication gap between the scientific researchers and industry (Lewin 2014).

Reaching a technical knowledge and a new product does not necessarily mean successful sales. Figure 9.8 shows numerous challenges to development and successful commercialization of nano-based products that can be categorized as:

1. Inadequate regulatory frame-work;
2. Lack of support and acceptance by the public, practicing physician, and industry;
3. Developmental considerations like scalability, reproducibility, characterization, quality control, and suitable translation;
4. Toxicological issues and safety profiles (immigration of nanoparticles from polymer matrix to environment);
5. Lack of available multidisciplinary platforms;
6. Poor intellectual property protection (Hobson 2009).

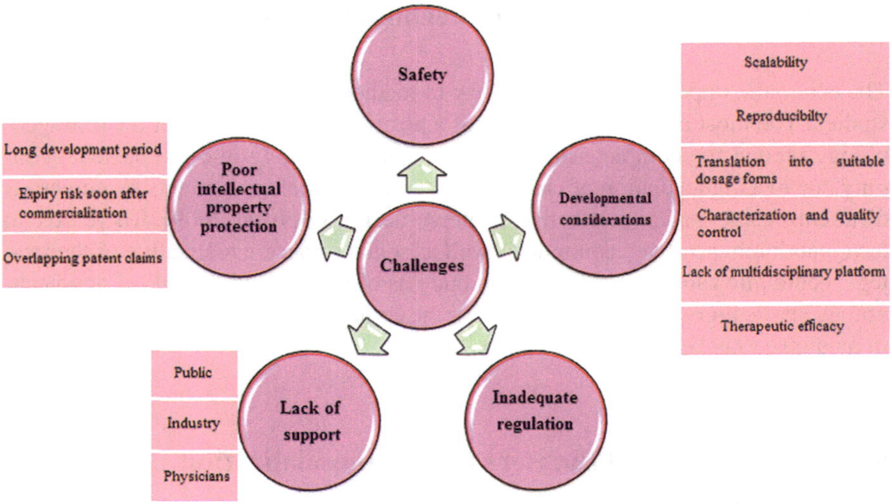

Fig. 9.8 Challenges to development and commercialization

Formulating an efficient and nontoxic product are the two key fundamentals for a product to successfully reach market. There are many challenges to development and commercialization of nanoproduct such as safety, developmental considerations, inadequate regulation, lack of support (industry, public and physicians), and poor intellectual property protection (Bastani and Fernandez 2002; Kaur et al. 2014). Overall, the basic research direction of the field, the biggest challenges are cost, the identification of commercial opportunities and lack of business skills in university researchers (Lewin 2014; Palmberg 2008). Market acceptance as barriers to commercialization can be difficult to overcome for obligatory firms (Lewin 2014).

There are investors' concerns such as selection between conventional products and nanotech-based products that are shown in Fig. 9.9. There are conventional products where yields on investment may be 3–5 years away, whereas the nanotech-based product has high and risk of delay or even low yields during the initial phase, and may require large investments (Kaur et al. 2014).

The limited impacts of many public sector ideas to endorse nanotechnology can be attributed to lack of ability and interest. The efficiency of the research and development, technical and training assistance, organized and economic support can be questioned since numerous ideas are not fully commercial.

9.9 Commercial Applications

Nanoparticulates, nanodevices, products containing or constructed using nanomaterials or containing nanostructures, nanomachines, and nanoelectronic components are commercial applications of nanotechnology. These developments are key

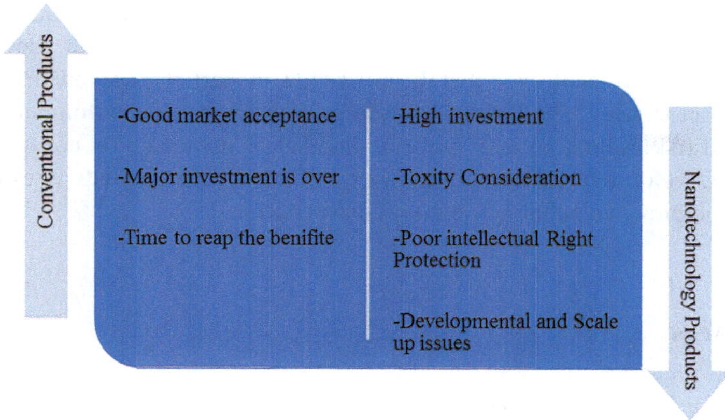

Fig. 9.9 Investors' concerns: selection between conventional products and nanoproducts

Table 9.4 Stages from basic nanomaterials to commercial applications

Stage	Examples		
Nano material or nanostructure	Nano-transistors	Copper nano-wires	Protein nanoparticles
Nano products	Improved semiconductor	Lithium-ion batteries	Carriers of drugs
Development of nanoproducts	Computers or small phones	More powerful battery for vehicles	Development of cancer cells of drug

drivers for the commercialization of all highly promising and safe medically-directed nanotechnologies. It is necessary that critical aspects of successful commercialization be realized early such as appropriately identifying and analyzing markets to a specific technology, developing project teams and plans in order to achieve the potential for new and developing nanotechnologies. For the development of any nanoproduct, proper attention on these critical elements is essential to succeed in commercialization (Hobson 2009; Mahdavinejad et al. 2014).

The application and commercialization of nanotechnologies are predictable to cut across established knowledge, technological and organizational boundaries and might disrupt traditional industries. This suggests that the ability to integrate knowledge, such as scientific, technological, commercial, regulatory, distributed across professional groups, companies and research organizations, is essential for nanotechnology development (Von Raesfeld et al. 2012). Applications include farming technologies and inputs, food processing, food packaging and retailing (Scrinis and Lyons 2007). Table 9.4 shows the transition from basic nanomaterials to commercial applications.

In this chapter we tried to introduce commercialization of nanotechnology and propose strategies for nanotechnology-based product marketing. Public concern grows with the increasing market trend of nanomaterial because of possible damage to human and environmental health. Many research studies have established the probable migration behavior of nanomaterials from the matrix, but some experimental studies have confirmed that the migrated nanoparticles are low in relation to other migration rates. The lack of knowledge on human and environmental health effects and risk evaluation may limit the number of nanomaterial consumption in industry. Developing, testing and analyzing of new nano-products is essential to develop and preserve public as investor confidence.

References

Balachandra P, Nathan HSK, Reddy BS. Commercialization of sustainable energy technologies. Renew Energy. 2010;35(8):1842–51.

Bastani B, Fernandez D. Intellectual property rights in nanotechnology. Thin Solid Films. 2002;420:472–7.

Bumbudsanpharoke N, Ko S. Nano-food packaging: an overview of market, migration research, and safety regulations. J Food Sci. 2015;80(5):R910–23.

Fiedler M, Welpe IM. Antecedents of cooperative commercialisation strategies of nanotechnology firms. Res Policy. 2010;39(3):400–10.

Frewer L, Gupta N, George S, Fischer A, Giles E, Coles D. Consumer attitudes towards nanotechnologies applied to food production. Trend Food Sci Technol. 2014;40(2):211–25.

Gómez Uranga M, Etxebarria Kerexeta G, Campàs-Velasco J. The dynamics of commercialization of scientific knowledge in biotechnology and nanotechnology. Europ Planning Studies. 2007;15(9):1199–214.

Hobson DW. Commercialization of nanotechnology. Wiley Interdiscip Rev Nanomed Nanobiotechnol. 2009;1(2):189–202.

Kaur IP, Kakkar V, Deol PK, Yadav M, Singh M, Sharma I. Issues and concerns in nanotech product development and its commercialization. J Control Release. 2014;193:51–62.

Khosravi N, Sadeghi M. The marketing strategy for successful product development performance in Iranian nanotechnology-based enterprises. In: Industrial Engineering and Engineering Management (IEEM), IEEE Int Conference. IEEE; 2014. p. 270–274.

Lee C-J, Lee S, Jhon MS, Shin J. Factors influencing nanotechnology commercialization: an empirical analysis of nanotechnology firms in South Korea. J Nanopar Res. 2013;15(2):1444.

Lewin PME. Technology transfer: the case of nanotechnology. 2014

Mahdavinejad M, Nazari M, Khazforoosh S. Commercialization strategies for industrial applications of nanomaterials in building construction. In: Advanced materials research. vol. 829. Trans Tech Publications; 2014. p. 879–883.

Meyyappan M. Commercialization of nanotechnology. MoffettField: MM Associates; 2007.

Nikulainen T. Identifying nanotechnological linkages in the Finnish economy–an explorative study. Technol Anal Strateg Manag. 2010;22(5):513–31.

Oriakhi CO. Commercialization of nanotechnologies. Cambridge: Massachusetts Institute of Technology; 2004.

Palmberg C. The transfer and commercialisation of nanotechnology: a comparative analysis of university and company researchers. J Technol Transfer. 2008;33(6):631–52.

Pomykalski P, Bakalarczyk S. Financing nanotechnology commercialization. In: 3rd International Conference NANOCON, Brno. 2011. p. 101–5.

Porter AL, Youtie J, Shapira P, Schoeneck DJ. Refining search terms for nanotechnology. J Nanopart Res. 2008;10(5):715–28.

Rasmussen B. Is the commercialisation of nanotechnology different? A case study approach. Innovation. 2007;9(1):62–78.

Roco MC, Bainbridge WS. Converging technologies for improving human performance: integrating from the nanoscale. J Nanopart Res. 2002;4(4):281–95.
Scrinis G, Lyons K. The emerging nano-corporate paradigm: nanotechnology and the transformation of nature, food and Agri-food systems. Int J Sociol Agri Food. 2007;15(2):22–44.
Shapira P, Wang J. From lab to market? Strategies and issues in the commercialization of nanotechnology in China. Asian Business Manag. 2009;8(4):461–89.
Tolfree D, Jackson MJ. Commercializing micro-nanotechnology products. Boca Raton, FL: CRC Press; 2007.
Tsuzuki T. Nanotechnol commercialization. Boca Raton, FL: CRC Press; 2013.
Von Raesfeld A, Geurts P, Jansen M, Boshuizen J, Luttge R. Influence of partner diversity on collaborative public R&D project outcomes: a study of application and commercialization of nanotechnologies in the Netherlands. Technovation. 2012;32(3):227–33.
Walsh PR. Innovation nirvana or innovation wasteland? Identifying commercialization strategies for small and medium renewable energy enterprises. Technovation. 2012;32(1):32–42.
Youtie J, Iacopetta M, Graham S. Assessing the nature of nanotechnology: can we uncover an emerging general purpose technology? J Technol Transfer. 2008;33(3):315–29.
Zucker LG, Darby MR. Socio-economic impact of nanoscale science: initial results and nanobank. NBER Working Papers 11181, National Bureau of Economic Research; 2005.

Chapter 10
Future Prospects of Nanobiotechnology

Nanotechnology and biotechnology enjoy a great deal of overlap in many research laboratories. Nanobiotechnology, as the name suggests, refers to the interface between, and convergence of, nano- and biotechnology. It is a multitude of various disciplines such as chemistry, physics, biology and engineering (Fakruddin et al. 2012). Main applications and opportunities of nanotechnology in the agri-food sector is shown in Fig. 10.1.

Nanoscale control of food molecules could allow the modification of many macroscale characteristics of foods such as texture, sensory attributes, processability and shelf life. The tools of nanotechnology have already allowed scientists to better understand the way in which food components are structured and how they interact with each other. This understanding is expected to enable a more precise manipulation of food molecules for the design of healthier, tastier and safer foods. Development of biosensors for pathogen and contaminant detection in food products helps ensure the safety of the food supply. Furthermore, conversion of renewable agricultural materials or food waste into energy and useful by-products is an environmentally oriented area of research that could be greatly enhanced by biotechnology (Moraru et al. 2009; Patel et al. 2018).

Nanomaterials can be utilized to make packaging that keeps the items inside fresher for longer. Insightful food packaging utilizing nanosensors could even furnish consumers with data on the condition of the food inside. Food packaging materials can be produced with nanoparticles that can alert buyers when an item is no longer safe to eat. Sensors can caution before the food product spoils or can tell us the precise nutritional status contained in the substance (Sundarraj et al. 2014). Embedding bioactive mixes and probiotic microscopic organisms inside prebiotic substances to ensure, or even increase, their survival while passing the upper gastrointestinal tract is a field of extraordinary enthusiasm for both the scholarly world and food businesses. An example of this innovation involves packaging bioactive mixes in micro- or nanoscaled particles that disconnect them and control their discharge until particular conditions arise. Electrospun nanofibres can likewise be utilized as the conveyance

H. Jafarizadeh-Malmiri et al., *Nanobiotechnology in Food: Concepts, Applications and Perspectives*, https://doi.org/10.1007/978-3-030-05846-3_10

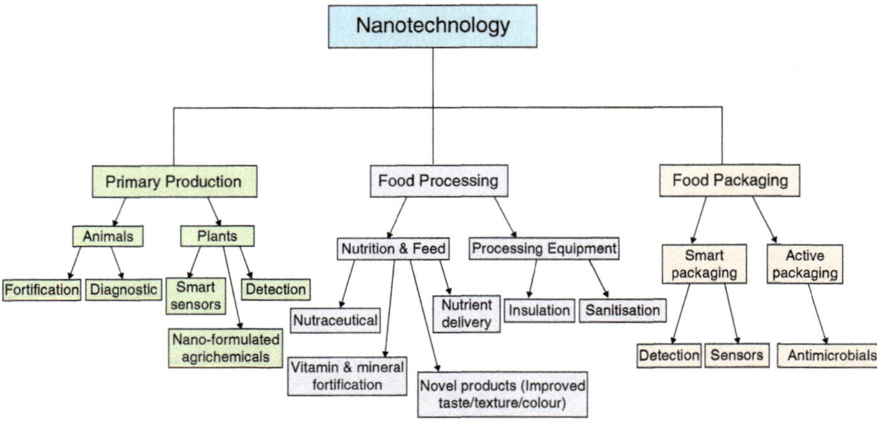

Fig. 10.1 Main applications and opportunities of nanotechnology in agri-food sector (Handford et al. 2014)

framework in food supplements to secure them amid handling and storage, or in conveyance frameworks for exchanging the segments to a distant site in the body. Finally, many sustenance-grade polymer frameworks (i.e. natural sustenance hydrocolloids) are difficult to create with electrospinning because of their poor viscoelastic conduct, absence of adequate atomic ensnarement, and constrained dissolvability. Generally, because just several handling parameters can be controlled straightforwardly as a portion of the included parameters are either exceptionally associated or obtained from the properties of the utilized polymer arrangement. Modern created electrospun nanofibres achieve a dimensional spread of the request of 10 rate focuses or worse among ostensibly identical specimens. Along these lines, a principle test related with the large-scale manufacturing of nanofiber fabrics is the execution of systems permitting improvements in the procedure and item reproducibility and to develop classes of utilizable materials (Chung et al. 2017; Gouveia 2010).

At present, the food industry is trying to identify opportunities and the most promising discoveries in nanobiotechnology that can significantly enhance food processing and food products. A limited number of applications that deal with matter at the nanoscale and are relevant to the food sector are currently available, and can be categorized in two groups: (1) functionalized membranes, which could be used to isolate and purify highly sensitive bioactive compounds and (2) nanostructuring technologies, such as emulsification, dispersion and nano-aeration, or the encapsulation of bioactive compounds in food polymer matrices. The fabrication of food biopolymer, based nano-structured materials with unique mechanical and functional properties, the synthesis of liposomes, which can be used for the encapsulation and targeted delivery of bioactive compounds through foods, or the development of nanosensors for pathogen detection, are all viewed as a starting point for the integration of nanobiotechnology into food science and food technology (Moraru et al. 2009; Weiss et al. 2006).

In summary, the main future prospects of nanobiotechnology in the food sector are

- health promotion through foods, by using food matrices as delivery tools for bioactive compounds
- development of ingredients (i.e. nanoadditive) with enhanced functionality, which could be used to manufacture foods with novel and unique flavors and textures
- development of novel, nano-structured packaging materials with enhanced barrier or self-sealing characteristics
- materials with microbial-repellant characteristics, which could be used to manufacture food contact surfaces with self-sanitizing properties
- novel processing technologies
- advanced tools (i.e. nanosensors) for food safety shelf life or quality monitoring
- enzyme immobilization using nanomaterials (i.e. organic and inorganic nanoparticles)

At present, a vast number of immobilization methods are currently available. Unfortunately, there is no universal enzyme support as the best method of immobilization might differ from enzyme to enzyme, from application to application and from carrier to carrier. Accordingly, the approaches currently used to design robust industrial immobilized enzymes are, without exception, labeled as 'irrational' because they often result from screening of several immobilized enzymes and are not designed. As a consequence, some of the industrial enzymes are working below their optimum conditions (Elnashar 2010; Vaghari et al. 2016).

References

Chung I-M, Rajakumar G, Gomathi T, Park S-K, Kim S-H, Thiruvengadam M. Nanotechnology for human food: advances and perspective. Front Life Sci. 2017;10(1):63–72.

Elnashar MM. Immobilized molecules using biomaterials and nanobiotechnology. J Biomater Nanobiotechnol. 2010;(1):1.

Fakruddin M, Hossain Z, Afroz H. Prospects and applications of nanobiotechnology: a medical perspective. J Nanobiotechnol. 2012;10(1):31.

Gouveia IC. Nanobiotechnology: a new strategy to develop non-toxic antimicrobial textiles. In: Mendez-Vilas A, editor. de current research, technology and education topics in applied microbiology and microbial biotechnology. Badajoz: Formatex Res Center; 2010. p. 407–14.

Handford CE, Dean M, Spence M, Henchion M, Elliott CT, Campbell K. Nanotechnology in the agri-food industry on the island of Ireland: applications, opportunities and challenges. 2014.

Moraru C, Huang Q, Takhistov P, Dogan H, Kokini J. Food nanotechnology: current developments and future prospects. In: Global issues in food science and technology. 2009. p. 369–399.

Patel A, Patra F, Shah N, Khedkar C. Application of nanotechnology in the food industry: present status and future prospects. In: Impact of nanoscience food industry. 2018. p. 1–27.

Sundarraj A, GuhanNath S, Aaron S, Ranganathan T. Recent innovations in nanotechnology in food processing and its various applications—a review. Int J Pharm Sci Rev Res. 2014;29(2):116–24.

Vaghari H, Jafarizadeh-Malmiri H, Mohammadlou M, Berenjian A, Anarjan N, Jafari N, Nasiri S. Application of magnetic nanoparticles in smart enzyme immobilization. Biotechnol Lett. 2016;38(2):223–33.

Weiss J, Takhistov P, McClements DJ. Functional materials in food nanotechnology. J Food Sci. 2006;71(9):R107–16.

MIX
Papier aus verantwortungsvollen Quellen
Paper from responsible sources
FSC® C105338

Printed by Books on Demand, Germany